STUDY SKILLS FOR GEOGRAPHY, EARTH AND ENVIRONMENTAL SCIENCE STUDENTS

This guide will help you to survive and thrive during your degree and on into the workplace. Everything you do at university can be useful in your career. Packed with practical hints, study tips, short cuts, real-life examples and careers advice, the new expanded fourth edition of this book is an invaluable resource throughout your geography, earth science or environmental science studies.

This book provides guidance for successful study on many topics including:

- Starting as a student
- Being an effective researcher
- Presenting information effectively in posters, presentations, essays and reports
- Time management, well-being and ethics
- Field and laboratory work
- Assessment and feedback

Written in an accessible style, this guide also explains the role of the academic, and how it differs from that of a school teacher. It prepares you for the world of work by showing how the skills you learn at university today can be used in your career choice of tomorrow.

Pauline E. Kneale is an Emeritus Professor at the University of Plymouth, UK.

D0162121

STUDY SKILLS FOR GEOGRAPHY, EARTH AND ENVIRONMENTAL SCIENCE STUDENTS

Fourth Edition

Pauline E. Kneale

LONDON AND NEW YORK

Fourth edition published 2019
by Routledge
2 Park Square, Milton Park, Abingdon, Oxon, OX14 4RN

and by Routledge
52 Vanderbilt Avenue, New York, NY 10017

Routledge is an imprint of the Taylor & Francis Group, an informa business

First edition published by Hodder Education 1999
Third edition published by Hodder Education 2011

British Library Cataloguing-in-Publication Data
A catalogue record for this book is available from the British Library

Library of Congress Cataloging-in-Publication Data
Names: Kneale, Pauline E. (Pauline Estner), 1954– author.
Title: Study skills for Geography, Earth and Environmental Science students / Pauline E. Kneale.
Description: Fourth Edition. | New York : Routledge, 2019. | "[Third edition published by Hodder Education 2011]"—T.p. verso. | Includes bibliographical references and index.
Identifiers: LCCN 2018051771| ISBN 9781138494121 (hardback : alk. paper) | ISBN 9781138494138 (paperback : alk. paper) | ISBN 9781351026451 (eBook)
Subjects: LCSH: Geography—Study and teaching (Higher)
Classification: LCC G73 .K547 2019 | DDC 910.71/1—dc23
LC record available at https://lccn.loc.gov/2018051771

ISBN: 978-1-138-49412-1 (hbk)
ISBN: 978-1-138-49413-8 (pbk)
ISBN: 978-1-351-02645-1 (ebk)

Typeset in Garamond and Helvetica
by codeMantra

Contents

Preface

This is the fourth edition of the Skills book designed for students studying Geography, Environmental Science, Earth Science (GEES) and related disciplines, especially those involving fieldwork: Transport Studies, Tourism, Ecology, Land Management, Landscape Studies, Marine and Ocean Studies. The book has been completely reorganised into five sections that should prove helpful! The order follows the learning cycle with transitional and general information at the start. Chapters on well-being and introducing university services support are new and upfront. Section 2 concentrates on the skills for learning which make Section 3 researching skills effective. Section 4 focuses on assessments and, finally, Section 5 covers fieldwork. The feedback from colleagues and students on previous editions has been included where possible – thanks for all the comments.

He just doesn't get my geology jokes

Thanks are offered to all the undergraduates and postgraduates that survived Study Skills modules and contributed ideas and prompts for this book. For this edition thanks are especially due to Wendy Miller, Sylvie Collins, Susie Stillwell, and Jane Dalrymple.

That's okay, igneous is bliss

Victor Gollancz Ltd gave permission to quote extracts from *Interesting Times* (1994) and *Maskerade* (1995) by Terry Pratchett.

This edition has two main messages:

KEEP SMILING!
and
JUST DO IT!

Section 1

Getting started, keeping going

Congratulations and welcome to the BOOK (or e-version!). Whatever your degree scheme: Geography, Environmental Science and Earth Sciences (GEES) or related disciplines such as Transport Studies, Tourism, Ecology, Marine and Ocean Sciences, Land Management and Landscape Studies, this book should help YOU to be successful. It should demystify aspects of university life and study, and build your confidence to learn effectively.

Section 1 focuses on information which will be particularly helpful at the start of your degree, and worthwhile throughout. Chapter 1 has essential information for when you start, and is worth reading later to see what might have passed you by in the excitement of the early weeks. 'Reflecting, reviewing and acting', Chapter 2, is designed to help you think through what is happening in university and other parts of your life, and make choices as you progress. It's about taking action, having control of your study. This links very closely to Chapter 3, a very practical approach to managing time, helping you to fit in many activities and really enjoy all aspects of university life. Chapters 4 and 5 cover useful stuff that may be vital depending on your circumstances. Skip-read the subheadings to see what is valuable for you. 'Looking after yourself', Chapter 4, should help you recognise some of the pressures of university life and point you to the huge range of supportive people and resources. The last chapter in this section concentrates on 'employability matters', because it does! GEES students are hugely employable, but you need the confidence to present your expertise in a way that employers understand and leads to job offers, and the huge range of jobs available to you is confusing. This chapter should enable you to make effective links between your university study, extracurricular and part-time work experience, and take time to explore opportunities, which will make your final year and transition to the workplace easier.

How to use this book

No single idea is going to make a magic difference to your learning, but taking time to think about (*reflect on*) the way you approach tasks like reading and thinking, listening and writing, researching and presenting should help. Studying is a personal activity. There are no 'right' ways, but there are tips, techniques, short cuts and long cuts. Reading this book in one go will not help. Look through the chapter headings and index. Read a couple of things that interest you now. When you are uncertain, worried or stuck at any point in your degree study, there should be a useful section. Some parts are relevant in the first year, others, like the dissertation advice (Chapter 25), will matter more in later years.

The **Try This** activities encourage you to get involved and build on your existing skills. Adapt the activities to your needs. Some statements in the book are deliberately controversial, designed to encourage thinking. Most of the figures and examples are deliberately 'less than perfect'. Consider how they can be improved; it's called *active criticism*. Universities have IT and sports facilities – getting more skilled means using the facilities and 'working out'. Your first year is a good time to practise and enhance your study skills as you adjust to your new life, but it is important to keep practising and reflecting throughout your degree. Experience is built by doing, not by watching.

Most people using skills books or websites, or on training courses, find that 90 per cent of the material is familiar, but new ideas make involvement worthwhile. The 90 per cent increases your confidence that you are on the right track, and the new ideas spark rethinking, reassessment and refining. You have been studying for 15 years at least, and are very skilled in various ways. Some approaches at university will be new, but a quick glance here should help you to understand what's going on. Chapters are deliberately very short, giving you the basics, some activities to practise and references to other texts and websites. Whole guides are written about essay writing, dissertations and fieldwork. Where you need more information, search online and at your university's learning skills centre.

The trick with university study is to find a combination of approaches that suit YOU, promote your research and learning, and enhance your self-confidence. As with all books, not all the answers are here, but who said study would be a doddle?

I hope that you will enjoy some of the humour; this book is meant to be gneiss and fun with serious points. Geologists probably have the best jokes, but they are all pretty dire. The crosswords follow the quick crossword style familiar to readers of UK newspapers; the answers have vaguely GEES connections. The fun activities give your brain a quick break. Light relief is vital in study; if you find deep thinking leads to deep kipping, have a coffee and solve a crossword clue – just remember to restart thinking after your break! Enjoy studying at university – it is exciting, fun and a challenge.

CARPE DIEM

'May the Force be with you' – Yoda

Wordsearch 1

Can you find the words and phrases in the grid? Words are written forwards, backwards and diagonally. Solution in Chapter 28.

Contour, Concept, Environment, Fern, Fohn, Friends, Geography, Geology, Heave, Idea, Icefall, Oasis, Online, Pipe, Research, Tea break, Tilt, Travel, Vacation, Workout, Woods.

F	R	I	E	N	D	S	D	O	O	W
T	G	B	R	R	L	D	I	G	E	O
P	E	V	A	E	H	E	E	S	N	R
E	O	A	A	F	S	O	V	L	A	K
C	L	N	B	C	G	E	I	A	R	O
N	O	L	I	R	A	N	A	T	R	U
O	G	N	A	D	E	T	B	R	L	T
C	Y	P	T	F	D	A	I	V	C	A
S	H	I	C	O	E	Q	K	O	E	H
Y	L	P	W	H	U	C	R	D	N	X
T	N	E	M	N	O	R	I	V	N	E

Transitions – essential info

'Teach? No,' said Granny. 'Ain't got the patience for teaching.
But I might let you learn.'

(Pratchett 1995)

So many new people – what do they do? Well, ask them – it's a great way to break the ice and find out about all the people that will be important in your life for the next … years. Amongst them you will find:

☺ Lots of new and experienced students – just like you. Friends for the next … years and some friends for life.

☺ Lecturers – they tutor, coach, mentor, facilitate, do administration, plan and deliver teaching, move classroom furniture about, troubleshoot problems and research when they can find time.

☺ Professors are more experienced lecturers with tutoring, teaching, administration and research to keep them busy. They are probably involved in university administration.

☺ Postgraduate tutors and demonstrators are students doing higher degrees and helping with teaching. They can be great fun, hugely helpful in class, practicals, on fieldwork and can tell you the best places to buy bread …

☺ School administrators are the really useful, lovely people in the School/Faculty Office who know everything, and they look after module and programme records, student records, assessment submission and can point you to all the other university services.

☺ Student representatives are elected by students each year to represent everyone's views on modules, programmes and the whole student experience. In university meetings they contribute to strategic discussions, policy development and enhancement of the university experience.

☺ Programme and module leaders – these are lecturers and professors who lead on the design and delivery of programmes and modules. They normally chair student–staff consultative committees each term, considering what did and didn't go well, and work on updating and enhancing your student experience.

Universities can be very big organisations with deans, heads of school, librarians, learning advisors, head of student services, residences staff … and many others – find out what they do. The Vice-Chancellor or President runs the university; you may go through three years without meeting him or her.

1.1 What are your priorities?

'*The problem with my first year was I didn't know what I didn't know, and even when I thought there was something I was supposed to know, I didn't know what to do about it*'. University can be confusing; it's not high school. The biggest change is that you are expected to **learn independently**, rather than being taught. This sounds scarier than it is. Talk to friends and family who have been to university. Ask how they found it compared with school. What insights do they add to the student comments below? What are you experiencing? What might you do in response?

University is not like high school	
Lectures have hundreds of people.	*The classes seem OK, I understand that much on the day, but there is so much more to get from reading and lab work.*
There are so many bright people.	*You have to organise yourself all the time.*
There were always so many other things to do, going out …	*At the start you don't think it's going to be too difficult, but if you want to do well it is much harder than school.*
Lots of people want you to do things with them.	*Mum isn't there to tell you what to do.*
Tutors expect you to talk about stuff … and think for yourself, it's not just about copying down the notes.	*I really wasted my first year, it was fun but it made the second year much harder.*

This activity involves talking to people to understand their experience (*research*), considering your own experience (*reflection*), deciding what is valuable for your purposes (*evaluation*), and doing something about it (*taking action*). It is a micro-scale example of the research process that is typical of higher education. If you talk to your colleagues about your results, reflections and planned activity then you are providing *feedback* to others and *disseminating* your research results through *discussion*. The topics explored in this book are selected because they are part of university life – taking a few moments to consider how you approach them will help you to be successful in your study and research activities.

University should be a good mix of fun, meeting lots of people and learning more about the subjects you have chosen to study. What do your friends consider to be the benefits of their university experience? Their replies might include:

- Having the chance to do interesting things;
- Having a better-paid, more interesting job;
- Having a more interesting life.

In my experience the people who do well and get the most from university keep a good balance between study and fun. Most people want to perform at a high standard in all assessments to get a good degree, knowing that writing and presenting at a high standard is a very useful life and workplace skill. They also want to be healthy and fit, and to develop good relationships with lots of people, and some people will become lifelong friends.

1.2 Learning opportunities – how most degrees are organised

Most undergraduate degrees in the UK involve three years of full-time study called either Years 1, 2 and 3, or Levels 1–3. An extra year may be intercalated (added) for an industrial placement or a year abroad. Higher education degrees are described as progressive, which means that the standards and difficulty increase each year. Expect modules at Levels 2 and 3 to build on your learning in earlier years (*keep your notes safe*).

A year is typically divided into 6 teaching modules or units, for a total of 120 credits. In UK universities the expectation is that each 20-credit module will occupy 200 hours of class, assessment and independent study time. It is usual for laboratory-based modules to have more class time than those involving lectures and seminars, and there is more class time in first year. **Your task** is to manage your time so each module gets equal attention.

In some universities you must pass all modules to progress to the next year. In other universities you can fail a module, but remember that all your results will be on your final transcript so your first employer will see these marks.

Your timetable may show the different learning activities. Expect to see lectures, seminars, tutorials, workshops, computer and science laboratories and fieldwork.

Lectures

Believing any of the following statements will seriously damage your learning from lectures:

- ✗ In good lectures, the lecturer speaks, the audience takes very rapid notes and silence reigns.
- ✗ The success of a lecture is all down to the lecturer.

✗ A great lecturer speaks slowly letting you take beautifully written, verbatim notes.

✗ Everything you need to know to get a first class degree will be mentioned in a lecture.

✗ Lectures are attended by students who work alone.

Lectures are the most traditional style of teaching, usually involving one lecturer and loads of students. Lectures involving 300+ students can seem impersonal and asking questions is difficult. A 20-credit module will probably have 40 hours of lectures but this is just part of the 200 hours you are expected to spend studying the topic. You can expect the lecturer will:

✓ Introduce him/herself, explain the objectives for the lecture and outline the learning outcomes, and be passionate and enthusiastic.

✓ Use videos, pictures, quizzes, games, role-play … to help you understand the material.

✓ Ask you to complete reading, or watch a video before the session as preparation and to enable discussion in class. This is called the flipped classroom approach.

✓ Talk for ten minutes and then ask you to think/discuss/debate/search for …

✓ Ask questions to see if you are understanding the topic.

✓ Include material and examples that would be difficult for you to find, and direct you to specific items to follow up through reading, discussion and group study activities.

✓ Provide opportunities to ask questions through and after the session.

✓ Use the last few minutes to summarise, to indicate what is coming in the next session and give you some questions or ideas to research in your reading. Capturing these points can be exceptionally helpful when revising.

✓ Provide advice on assignments for the module, answers to in-class tests, and feedback on assignments.

A lecturer is aiming to introduce the topic so that you will rush to read in more depth and really understand the topics that interest you. To manage learning from lectures:

☺ Download and read the PowerPoints for the lecture from the VLE. In many universities they are available up to a week ahead.

☺ Get to the lecture theatre early; find a seat where you can see and hear.

☺ Have your notepad and pens ready for each session if writing is your preferred note-making method. Make sure your laptop or tablet is fully charged for typing notes, and take paper and pens for emergencies.

☺ Get your brain in gear by thinking, '*I know I will enjoy this lecture, it will be good, I really want to know about …*'; '*Last week s/he discussed …, now I want to find out about …*'

☺ Before a lecture, read the notes from the last session. Skim-read (Chapter 10.2) the PowerPoints. Just 5–10 minutes will get a fuzzy brain in gear. If you know roughly what is being covered you remember more. Making notes on PowerPoints saves you writing down the main points – you can add other details.

☺ Look at handouts carefully. Many lecturers provide summary sheets with lecture outlines, main points, diagrams and reading, either hard copy or on the module website. Use these to plan reading, revision and preparation for the next session.

☺ Think critically about the material presented. What did you not understand? What do you need to read to get more case examples? What is interesting? Why is this topic important?

☺ Revise and summarise notes soon after a lecture. It helps revision later on and you can see where there are gaps in your understanding, which helps you to decide what to read.

☺ Talk about the topic with friends, study buddies and others.

If your university records all the teaching sessions, you have the opportunity to catch up later. Working from these recordings is a good second-best, but you will miss the group discussions and activities which really fix ideas in your head.

Skills acquired during lectures include recognising new information and identifying where you need to do further work, recognising research frontiers and subject limitations.

TRY THIS 1.1 – My lecture habits

Look at the three sets of bullet points above (✗ ✓ ☺), and reflect on what you did in your last lecture, picking two points to work on for your next lecture. Ask yourself whether this change helps you? Try it again on something else. You decide.

(Hint: Don't be too ambitious – trying one thing and thinking about whether it helps, and then trying something else, allows you to critically evaluate whether the change is successful for you.)

Tutorials: What are they? What do you do?

Usually tutorials involve a 50-minute discussion meeting with an academic or postgraduate chairperson and 4–8 students. The style varies between departments, but expect to prepare something in advance, typically a short talk, essay, an outline essay, points to debate, a computer programme, map, flow diagram, review of a paper … and to share your information. The aim is to discuss and evaluate issues in a group that is small enough for everyone to take part.

Other jolly tutorial activities include brainstorming examination answers, working through maths or statistics problems, comparing note-making styles, creating a poster, planning a research strategy, discussing the practicalities of a fieldwork proposal, evaluating dissertation possibilities, and the list goes on.

The tutor's role is NOT to talk throughout, NOT to teach and NOT to dominate the discussion. A good tutor will set the topic and style for the session well in advance, so everyone knows what they are doing. S/he will let a discussion flow, watch the time, make sure everyone gets a fair share of the conversation, assist when the group is stuck, and sum up if there isn't a student summariser role as part of the activity. A good tutor will comment on your activities, but tutorials are YOUR time.

Expect a tutor to ask you to run a couple of tutorials in their absence, and to report a summary of the outcomes. This is not because tutors are lazy, but because generating independence is an important part of university training. Student-led and student-managed tutorials demonstrate skills in the management of group and personal work. When a tutor is ill, carrying on the session without your tutor uses the normal tutorial time effectively. A tutor may assign, or ask for volunteer chairpersons, timekeepers and reporters to manage and document discussions.

Your role is to arrive at tutorials fully prepared to discuss the topic NO MATTER HOW UNINTERESTED YOU ARE. Use tutorials to develop listening and discussion skills, become familiar with talking around GEES issues and build up your experience of arguing about ideas. Tutorial skills

> *I don't really like tutorials because they make you think. You cannot just sit there. I do know that listening and talking about stuff makes me think, and that gives me ideas.*

⇨ TOP TIPS

- ✓ Taking time to prepare for tutorials will stop (or reduce) nerves, and you will learn more by understanding a little about the topic in advance.
- ✓ Reviewing notes and reading related material will increase your confidence in discussions.
- ✓ Asking questions is a good way of getting involved without having to know the answer.
- ✓ Prepare a couple of questions or points in advance and use them early on – get involved.
- ✓ Taking notes in tutorials is vital. Other people's views, especially when different from yours, broaden your ideas about a topic, but they are impossible to recall later unless noted at the time. **Tutorial notes make good revision material**.

include communication, presentation, critical reasoning, analysis, synthesis, networking and negotiation.

Seminars

Seminars are a slightly more formal version of a tutorial, with 8–25 people. One or more people make a short presentation, leaving ample time for group discussion. Seminars are a great opportunity to brainstorm, to note the ideas and attitudes of colleagues, to spot extra examples and approaches. Take notes.

Even if you are not a main speaker, you need to prepare in advance. In the week you speak, you will be enormously grateful to everyone who contributes to the discussion. To benefit from this kind of co-operation, prepare and contribute in the weeks when you are not the main presenter.

ONLY TO BE READ BY THE NERVOUS: (*Thank you.*) If you are worried, nervous or terrified, then volunteer to do an early seminar. It gets it out of the way before someone else does something brilliant (well, moderately reasonable) and upsets you! This will strengthen your skills and build your confidence so seminars become less of a nightmare. By the third week you will know most people in the group, and be less worried.

Skills developed in seminars include discussion, listening, analysis, teamwork, giving a professional presentation and networking.

Workshops or large group tutorials

Workshops (large group tutorials, support or revision classes) are generally sessions with 12–35 students that support lecture and practical modules. They have very varied formats. There will usually be preparation work and a group activity. Tutors act as facilitators, not as teachers. Expect a tutor to break large groups into subgroups for brainstorming and discussions. These sessions present a great opportunity to widen your circle of friends and find colleagues with similar or different views, *and* develop discussion, argument and listening skills.

Workshop skills are the same as for tutorials and seminars, with wider networking and listening opportunities.

Fieldwork, computer and laboratory work

Practical classes and fieldwork provide multiple opportunities for 'hands-on' skills development in GEES degrees. Many departments assign practical class time, when tutor support is available, BUT completing exercises and developing your proficiency in IT, computing and laboratory skills will take additional time. Check the opening hours of computer laboratories on campus.

In laboratory practicals and fieldwork, always take note of safety advice, wear lab coats and safety glasses as advised, and please don't mix acids without supervision.

Laboratory and field course staff are trained in first aid, but would rather not practise on you.

Library, empty classrooms, cafés

Universities have many spaces where students learn independently. The library is the obvious learning environment, providing quiet areas and areas for group work and discussion. Informal spaces include cafés, seating areas, empty classrooms and rooms in halls of residence. Each student finds their preferred spaces for learning. Lounging on the library beanbags suits some people, others need continuous access to coffee. Sockets for recharging devices can be crucial. What suits you best for working alone and for learning with others? Explore different places on campus and try them out.

Students who commute for an hour or more each day report finding useful reading and discussion opportunities on trains (Chapter 4.4). Discussions here may involve students from other programmes and years, providing insights that campus-based students may miss.

1.3 Introducing learning!

Independent learning – what do they mean?

University is different from school. Lecturers will help, but the responsibility for learning lies with each student. This is called independent or autonomous learning. It allows people of different ages, backgrounds and interests to study together and graduate with degrees in the same subject, but each person will have explored their own unique combination of topics (Figure 1.1).

Effectively university is about taking personal control of what you do and how you do it. There are modules, fieldwork visits, laboratories and the time to explore many avenues. If, in the process, these equip you for later life, that is a bonus. GEES degrees have two elements:

- **The knowledge.** This involves current information, data and theories, from cultural and medical geography to the racing speed of warm-based glaciers, fossil forests in Antarctica, the consequences of deforestation, geopolitics, tectonics, sustainability and enterprise. The range and scope of GEES studies is planet-wide and deep.
- **The skills.** You will practise and polish an excellent set of practical, personal and interpersonal skills that are useful throughout life.

Most university websites include the 'graduate attributes' that students acquire through practical experience or osmosis during their degree. Search for Graduate

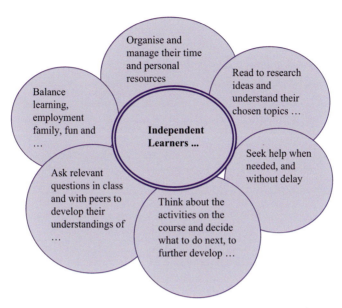

Figure 1.1 How independent are you? Which of these 'petals' needs your attention?

Attributes on your university website or generally. Attributes such as global perspectives, citizenship, sustainability awareness, creative and critical thinking ... may seem challenging, but these themes are likely to appear in your programme or in extracurricular opportunities.

Learning how to learn – lifelong learning

Most higher education involves active learning in environments where there is interaction, discussion and collaboration. Communication is crucial to re-evaluate and create ideas. Self-motivated learning is a vital skill for life, enabling you to keep abreast of developments and changes. Graduates from the 1960s, before computers and the internet, have had to get to grips with technology while at work; what will happen in the next 25 years? Employment is unpredictable. Job markets and business requirements change rapidly. Employers need individuals who are thoughtful and flexible about their careers. An effective graduate is someone who sees their career as a process of work and learning, mixing them to extend their skills and experience. This is the essence of lifelong learning, and university is part of the process.

To add value to your degree, it helps to think about (*reflect* on) what you do every day at university that gives you valuable workplace skills. Employers claim to be happy with the academic skills students acquire, such as researching, collating and synthesising new material, but they also want graduates with listening, negotiating and presenting skills. Any strengthening of your skills, and experience of skill-based activities, should add to your self-confidence and improve your performance as a student and as a potential employee. Expect to be involved in tutorials and activities that include:

Research	Personal development planning
Practical work	Action planning
Fieldwork`	Simulations
Role-plays	Case studies
Negotiation	Work placements
Interviews	Self-evaluation and review
Self-guided study	Group work
Learning contracts	Time management
Decision-making	Presentations

Your degree lets you experience the latest wikis, blogs, podcasts, vodcasts, electronic books and journals, video-conferencing, email, spreadsheets, e-learning, and VLEs. Some technology may seem daunting but it is fun too. (And if some 5-year-old proto-geek can manage, so can you!)

Q. Isn't the skills part of it just the stuff we had to do at school again?

A. No, some of the topics will be the same (presentations …), but at uni it is focused on behaving like a professional, developing a calm, polished and confident approach in all situations.

University teachers know that effective graduates are comfortable and confident with knowledge from their degrees (petrology, economics, air pollution) and the ways in which they acquire and share information and data (by individual research, group discussions, seminars, presentations, posters, reports). Good teachers encourage you to build on your experience through class activities which involve working with others as well as alone, because the reality of work and research is that they are team games. By the time you graduate you should feel confident in listing many skills on your curriculum vitae (CV), and be able to explain where in your degree these abilities were practised. Use **Try This 1.2 Skills this year**.

> **TRY THIS 1.2 – Skills this year**
> Take a look at your module information. Which skills are being developed? Where do they fit into this matrix and what else is involved?

Graduate skills	Hint
Learning how to learn	Taking personal responsibility for your learning as an individual and in group research for fieldwork, laboratory work, projects and dissertations.
Communication	Presentations in seminars, tutorials, workshops, debates, practicals and all written assignments.
Information Technology	Modules involving online research, data analysis, word processing, graphs and visualisations, statistical analysis, GIS and programming.
Numbers, data, numeracy	Modules involving statistics, data handling, analysis of data in practicals, fieldwork, projects and dissertations.
Mapping	Particularly important for earth scientists.
Professionalism	Presentations, report writing, discussions.

Assessment is a learning activity

Assessment comes like Christmas presents, regularly and in all sorts of shapes. Assessments should be regarded as helpful, because they develop your understanding. There are two forms:

✓ Module assessments where the marks do not count (formative), which usually involves lots of *feedback*.

✓ Assessments where the marks do count (summative). The results eventually appear on your degree result notification for the edification of your first employer who wants written confirmation of your university prowess.

There is a slight tendency for the average student to pay less attention to formative module assessments, where the marks do not count. Staff design formative tests because they know 99 per cent (± 1 per cent) of students need an opportunity to 'relax and have a go', to understand procedures and what is expected when the marks do not matter.

Assessments: examinations, essays, oral presentations, seminar presentations, posters, discussion contributions, debates, reports, reviews of books and papers, project designs, critical learning log, fieldwork notebooks, laboratory competence, computer-based practicals, multiple-choice tests … It is a matter of time

Q. *Why should I bother with peer assessment?*

A. *Peer assessment helps you to develop your critical assessment skills and become more comfortable with giving and receiving feedback. No staff, no stress!*

management to organise your life and them (Chapter 3). You should know in advance exactly how each module is assessed and what each element is worth. Modules have mixed assessments, so those who do very well in examinations or essay writing are not consistently advantaged.

Many departments have assessment and feedback criteria. Find them (!) or see examples for essays (Figure 18.2), presentations (Figure 22.2), practical reports (Figure 19.2), dissertations (Figure 25.2) and posters (Figure 23.3). If you cover all the criteria then the marks come rolling in.

Amongst the many skills enhanced by assessments are thinking, synthesis, evaluation, originality and communication.

Feedback is all about learning

Students receive and give feedback all the time; from tutors, lecturers, study buddies, friends, and everyone else who has their interests at heart. Lecturers provide feedback in conversations, by email and in comments on assignments because they want all their students to learn effectively and they use feedback as a form of coaching. They use many different feedback styles to encourage you to reflect and learn.

> *Feedback is everywhere. People talk to you about stuff … tutors, mates, people in hall. Feedback from peers and supervisors improves what you are doing. I had to work out how to use it so it was useful for me.*

Use **Try This 1.3** to make notes about how you respond to feedback positively (*… great idea, I am doing … now*) and negatively (*Don't you tell me what to do… grumble*), and how you can make the most of the advice offered in future.

⚙ TRY THIS **1.3 – Handling feedback**

Feedback arrives in many ways. How do you respond to each style to make the most of the advice? Make some notes. Try something out.

Verbal feedback from a lecturer in a large class/lecture.	Comments and ideas from a study buddy.	Text message from a tutor.
Online discussion using Skype, tutor to student, and students in peer groups.	Written feedback on a presentation, report or essay.	Verbal feedback podcast sent by mobile phone and VLE within 24 hours.
Conversation with your tutor in the corridor, refectory or gym.	Video clip or podcast from the marker which arrives three weeks later.	Summary written feedback on a class exercise online in the VLE/handed out in class.
Advice in module handbooks. This may include feedback on students' work in previous years.	Peer tutors (students mentoring students).	Information from students in the year above you. (It's feedback.)

Advice, comments, thoughts, information, opinions, suggestions, guidance, guidelines, instructions, recommendations, views, perspectives … these are ALL forms of feedback. Listening carefully is often the vital skill here: take some notes. Feedback is designed to enhance your skills and performance. Comments are made about your work, not about you as a person.

> Take feedback seriously, think carefully. If you disagree with some of it, it may be best to discuss why with your tutor. When advice is balanced and constructive it is probably useful. Discuss feedback with your friends and tutor.

Where classes are very large feedback may happen some weeks after you finished an assignment. It is too easy to glance at the mark and ignore your tutor's detailed comments. Work out a strategy to make sure you benefit from this feedback to improve your next assignment. Handling feedback so that you learn from it is a major university skill that has real workplace value.

Extracurricular experiences and learning

Do not underestimate what you know! In your years at university you acquire loads of personal skills: negotiating with landlords, debt crisis management, charming bank staff, juggling time to keep a term-time job and deliver essays to deadline, being flexible over who does the washing up, and handling flatmates and tutors.

While tutors are primarily concerned with your degree programme learning they will also stress the advantages of getting involved in extracurricular activities which make life more interesting and balanced. Universities are a great place to get involved with a new sport, hobby or volunteering. Your Students Union has many societies with friendly people thrilled that you are getting involved. Society participation is particularly helpful if you plan to develop your leadership, committee and organisational skills.

Study buddies and Peer Assisted Learning

Your tutors know that students work well in self-supporting study groups, sharing ideas and working on projects together. Group activities provide 'safe spaces' to try out ideas, to find what other people are reading, and to develop your confidence in discussions. You can expect opportunities for networking (getting to know people on your course) from the first week. Maybe there is someone you will work with? Buddying to share the cost of books, discuss assignments and have fun with makes sense.

Peer Assisted Learning (PAL) is offered in some universities, where students in more senior years run support activities, particularly for first-year students. PAL sessions (they may have a different name) are great places to improve your notes from a lecture, ask questions that are too difficult to ask in a room with 300 people, and plan what you want to read to consolidate your learning. Take advantage of the thinking time. You will also meet students from other years on your course: useful contacts. See West *et al.* 2017 for more details.

1.4 Why are there so many ways of learning?

University staff aim to develop your skills and experience, building your confidence through your degree. They also know that reading and hearing are the worst ways to learn anything. Stop reading and check out **Try This 1.4.**

> ⚙ TRY THIS **1.4 – Active or passive learning?**
>
> Search online for images of 'Cone of Learning' and 'Cone of Experience'. There are many to choose from. What are your thoughts (reflections)?

Now you know that you will have forgotten 90 per cent of everything you hear or read within two weeks, and that effective learning is active learning ... **how will you learn**? Small-group discussions in lectures are designed to stop passive listening and start active learning. Look back at Chapter 1.3 and sort out what is active and what is passive.

Lecturers have many roles: teacher, facilitator, advisor, assessor ... The very best are those who let you work stuff out for yourself, who don't tell you the answer to the question. These people are being very helpful in building your confidence to be an independent researcher (although annoying at the time).

1.5 Moving to university worries

It is a big step, exciting and worrying for some. Many universities have web pages to help you and answer your questions. Useful search terms include: moving to university; tips for freshers; welcome week; transition support; transition issues; university life-cycle; preparing for university. Most universities, programmes and halls of residence have their own social media spaces which let you connect before arrival, including with current students who usually provide good advice. Get involved, but don't put photos or comments online that you will regret later. (BTW – This section is very short because most readers have already arrived; there is lots of advice online).

We didn't do this at school

An increasing number of students study GEES degrees without substantial high school work. Earth Science students often study sciences at school, but little geology. This is not a disadvantage: you are interested in the subject, all topics are fresh, and you are not confused by half-remembered notes. Geography, earth and environmental sciences are all big and diverse subjects; no one gets to do all of it, at school or anywhere else. No problem. All three disciplines have traditionally recruited experts and ideas from other subjects.

> ☞ TOP TIPS
>
> ☺ Don't panic.
> ☺ Go to the lectures and get the general idea of the topic and material. Sort out the focus of the module. Plan your reading strategy.
> ☺ Pre-university: try local libraries for high school texts to pick out the background before a course starts, BUT you must move on to university-level texts. School texts are starting, not end, points.
> ☺ Take school notes to university; your soil notes may help someone who can help you via their economics notes. People with a maths background are very popular in statistics modules.
> ☺ If you have difficulty with essay writing, and especially if you have not written an essay in three years, find the student learning support team (ask in the library).
> ☺ Ask people which reading they find accessible and comprehensible.

1.6 What do lecturers mean by 'your own research'?

All students are 'reading' for a degree, finding out for themselves their own research process. Taught activities involve 20–50 per cent of the timetable, leaving 80–50 per cent for your research into sub-topics through activities that reinforce your understanding. The scope of GEES topics is vast, certainly global and at times interplanetary. Covering all aspects is impossible. Your challenge is to develop your interests and knowledge through researching different elements of the topics involved in your degree, within the constraints of time, facilities and energy. Consequently, researching in a group (see Chapter 16) can be hugely beneficial.

The issues addressed in the GEES degrees are not simple. Lecturers indicate what is already well understood and point out where knowledge is missing, provisional, uncertain and worthy of further investigation. Most issues are interlocking and multidimensional. By the time you graduate you should be able to take complex, unclear and, at times, contradictory information from a wide range of sources and synthesise it to make sense of the picture at a range of scales. For a number of topics which you have studied in depth you will have an enhanced ability to recognise both the boundaries of knowledge – what is known and what is not known – and what you as an individual know and do not know. Recognising the boundaries of one's own expertise is a life skill. Someone who does not understand the implications of their actions in changing procedures, for example, is a potential danger to themselves and the wider community.

> *University learning is not about recalling a set of lecture notes. It is about understanding issues and being able to relate and apply them in different contexts.*

1.7 References and resources

There are many generic skills texts – check out the library and online. For introductions to your discipline:

Daniels P, Bradshaw M, Shaw J, Sidaway J and Hall T 2016 *An Introduction to Human Geography*, (5th Edn.), Pearson, London.

Gregory KJ, Simmons IG, Brazel AJ, Day JW, Keller EA, Sylvester AG and Yanez-Aranciba A 2009 *Environmental Sciences: a student's companion*, Sage, London.

Holden J 2017 An Introduction to Physical Geography and the Environment, (4th Edn.), Pearson, London.

Pratchett T 1995 *Maskerade*, Victor Gollancz, London.

Rogers A and Viles H (Eds.) 2007 *The Student's Companion to Geography*, (2nd Edn.), Wiley-Blackwell, Oxford.

StudentGuide 2013 The Complete Guide to Geology, http://www.studentguide. org/the-complete-guide-to-geology/ Accessed 15 January 2019.

West H, Jenkins R and Hill J 2017 Becoming an effective Peer Assisted Learning (PAL) Leader, *Journal of Geography in Higher Education*, 41, 3, 459–465.

Geo-paths 1

Can you find six or more words in the square? Move horizontally and vertically, but not diagonally, from letter to touching letter. A letter can only be used once in a single word. One word uses all the letters. Answers in Chapter 28.

C	I	N	T
R	M	E	M
O	V	I	N
E	N	R	O

Reflecting, reviewing and acting – adding value to your degree

Experience is something you acquire just after you need it.

This is probably the most useful chapter in this book because throughout your degree you decide what to read, research, ignore, practise, and … (panic about). Being effective in reflecting (thinking and making notes), reviewing options (reading-list choices) and then acting will build your confidence and provide rational options when there are choices. Taking control and responsibility for your work, being an Active Learner (Chapter 1.4), and acting on feedback can seem scary, but it is core to independent learning, and the point of higher education.

The absence of detailed guidance about your studies causes some students to feel disoriented, chucked in at the deep end without a lifeguard in sight. Tutors will prompt you and make suggestions, but they are encouraging you to take decisions and gain the confidence to explore and think for yourself. Excellent tutors never tell you what to do. They will not micro-manage your learning. You attend modules with a hundred other students, but you each learn different things because you choose different readings and present different examples and ideas in assignments.

Actively managing yourself really helps you cope with life, people … and assignment deadlines. Reflection will help you see how different modules interconnect and how you develop skills as you progress. Taking a little time to think about interconnections between modules, 'why economic costing and pricing principles are important for cultural geographers or flood forecasters', or 'where statistical tests can be used in your dissertation', can give modules more cohesion and be motivating.

There are a number of **Try This** activities in this chapter to help you become familiar with the language and practice of reflection. Use a 'mental and physical' (fingers on the keyboard, record on phone, pen in hand) approach. The knack is to develop reflective thinking so that it becomes automatic, part of normal life. The **Try This** activities can be used to build up a *learning log*, recording your university experiences. Recording provides you with a diary, journal or collection of notes (phone notes, memory box, pc …) of your activities and skills, but that is just the first stage. The important distinction is that reflection and a *reflective log* goes beyond the diary element. It asks you to think (reflect) on what you did; decide what else you could have done (instead of, or as well as); decide what to do next; and talk about it fluently (self-advocacy). This is *active reflection*, giving you practice in developing more detailed, reasoned responses to what has happened and making changes.

Reflection skills are not learned easily or acquired overnight. They develop with practice; the result of thinking actively about experiences and placing them in a personal context. This is an iterative process, and a lifelong skill. Interestingly, most people find that the quality and value of reflection tends to be better when one is doing nothing in particular – washing hair, walking, waiting for someone – the trick is to capture the thoughts that you have and ask 'what else?'

2.1 Getting started

Sometimes you have to step back to leap forward.

Most universities have processes to help students to plan and reflect on their progress during their degree, using an e-portfolio or paper-based PDP (personal development planning) system. This structured format for planning and record keeping is likely to be introduced and used in tutorials and be part of some module assignments. The systems have various names usually involving a combination of: personal development, planner, portfolio, e-portfolio, progress file, learning-log and reflection. The advantage of e-portfolio PDP systems is that you can easily import pictures and video, and examples of your work from word processing files. During group work you can share feedback and thoughts more easily than in paper-based systems. Your programme

> *For most of my degree we did lots of different modules and they were interesting in a way, but I couldn't see why we were doing economic geography and then statistics and the resources bits and so on. I couldn't really see the point of working at the bits. It was about half way through final year when all the bits began to make sense and slot together. And then it got to be really interesting and I could see why I should be doing more reading and I did quite well in the last semester.*

will introduce you to your university's system; PebblePad (2018) is widely used. It provides a personal learning space where you can create, record and store your learning and personal and professional development information. The items you create or upload are private until you choose to share them. It is usual to share information as part of group work, and to share summaries of your reflections with tutors. *What you share is controlled by you.*

If you find lecturers use PDP in first-year modules, but not in second- or third-year modules, it's because they are familiarising you with the process in first year expecting that you will continue automatically, by yourself, later in your degree.

Some businesses require new staff to keep a daily log in their early years of employment. It encourages employees to assess the relative importance of tasks and to be efficient managers of their time. In the last ten minutes each day, you complete a statement like:

> I have contributed to the organisation's success/profits today by ....
> I was fully skilled to ....
> I was less capable at ....
> Other comments ....

At the end of the week or month, these statements are used to prioritise business planning and one's Continuing Professional Development (CPD). It is an activity that employees hate if they have no previous reflection experience. However, most admit that it taught them an enormous amount about their time and personal management style and wished they had started sooner. Practice enables this structured self-reflection to become automatic: individuals continually evaluate their personal performance and respond accordingly. How will you complete the student equivalent for yesterday, today and tomorrow?

> I contributed to my degree today by ....
> I could have been more efficient if I had ....
> Tomorrow I am going to ....

Developing the language of reflection

Active engagement with reflection increases your confidence in:
☺ deciding on your learning and career aims, evaluating opportunities and recognising difficulties, making choices about what and how to study, career and personal options;
☺ thinking through your values and beliefs, evaluating your current capabilities and what motivates you to make changes;
☺ setting aside time to reflect, pause, be introspective, your own 'critical friend'. These processes provide the opportunities to reconsider goals and achievements, and to reset the clock or take new directions in a thoughtful and reasoned way;

☺ articulating your achievements and setting these in the context of your personal and career development. Being confident and comfortable talking about yourself (self-advocacy) to different audiences, family, friends, employers, tutors …, allows you to critically examine and evaluate your activities.

Reflection takes practice. It involves a continuous loop or cycle of: action ➔ reflection ➔ decisions about next steps ➔ action ➔

Most people need to go through the cycle a number of times before they see significant benefits because thinking takes energy, so don't give up. Putting time in your diary for thinking *is important* – time when you won't be interrupted.

> *My first year tutor got us to fill in the forms, and you do get better by doing the practicals and some exercises. My second year tutor made everyone pick a skill to focus on each term. Having decided to get better at chairing, I volunteered to chair each of our group work activities and I wouldn't have done it if he hadn't made me pick one thing. It felt crazy weird because I don't normally volunteer but it was good, and I liked it after a bit.*

Try This 2.1 asks you to articulate your feelings about your personal approach to learning and your degree course, with some students' examples to help you get started. When you evaluate your responses and act on them, you are taking charge.

TRY THIS 2.1 – Initial reflections

Here are some example reflections – What are your thoughts? Make notes as you think. Notes may be hand written, electronic or a voice recording; which suits you best for planning?

Skill	Example reflections
Speaking in tutorials and seminars	*I know I don't say enough in tutorial. I know what I want to say but it all seems so obvious I feel silly, so I guess I need to get stuck in early. PLAN: to answer first, learning log can act just as a diary, somewhere to write down questions to ask.*
Knowing when to stop reading	*I read up to the last minute, and then rush the writing. PLAN: to put STOP READING deadlines in my timetable.*
Including relevant information in essays	*I include everything I can in an essay to show I have done lots of reading. PLAN: be more selective, somehow, next time.*
Organising ideas coherently	*I know I can do this if I plan an essay properly, getting paragraph ideas in order. PLAN: just do it next time.*
Being more open to new ideas	*PLAN: Find different news sources. Take notes in tutorial. More reading might be okay!*

| Making time to sit down and think about different ideas | *This seems really odd, because you sort of do thinking all the time. It's not really a cool activity. Could try when no one knows – Sunday morning maybe. Bath and jogging doesn't work, making notes is too difficult.* |

What are your thoughts on:
Making notes
Listening carefully in discussion and responding
Completing essays on time
Using diagrams to illustrate essay ideas
Drawing the threads of an argument together to develop a logical conclusion
Negotiating
Putting ideas into my own words
Disagreeing in the discussion without causing upset/being upset
Reading more widely
What else?

Reflection does make you more organised and independent. It takes practice.

Try This 2.1 is a self-assessment exercise that you could repeat after a term or semester. It is useful to remember that when people self-assess a skill before and after an activity, the assessment at the end is frequently lower than that at the start. Although the skill has been used and improved during the activity, by the end, it is possible to see how further practice and experience will lead to a higher skill or competence level. Now look at **Try This 2.2**. In thinking about your strengths and weaknesses, talk to family and friends, ask what skills you have and what you do well. Notice how they talk about you; do they use phrases you can adopt?

⚙ **TRY THIS** 2.2 – Being an active reflector
Take your *Try This 2.1* list and highlight three skills you would like to develop in the *next three months*. Make some notes about what to do about these three issues, add some activity deadline dates. Like New Year resolutions, this activity benefits from practice.

How do you make decisions? Everyone is different and there is no right way, but being aware of your approach can help you research and plan. Use **Try This 2.3.** You will probably ✓ a number of approaches depending on the circumstances, but overall *are you in control of your life*? Are you making your decisions? What next?

⚙️ **TRY THIS** 2.3 – Decision-making

Consider the decisions you have made in the last week and last five years. Considering your approach, ✓ the styles of decision-making that apply to you. Then make some notes about who really decided, consider whether you are comfortable with your approach, and what you might want to change.

Style	Explanation
Hopeful	Choosing the option that should bring a happy result.
Go for it	Get straight on with the first option that comes to mind, without considering any other paths.
Tomorrow will do	Leaving decisions until well past critical times, putting things off.
Alternatives are overwhelming	Researching thoroughly, but getting so much information that you cannot decide on priorities. (Cannot see the wood for the trees.) Prevarication takes over.
Following the crowd	Letting the group or another person (friends, family) do the thinking and deciding for you.
Fatalistic	Drifting, letting life happen, not being prepared for potential eventualities. Essentially not deciding.
Missing out/ Avoidance	Taking yourself elsewhere so that your imagined 'worst case scenario' cannot happen.
Risk averse	Taking the safest, best-odds opportunities.
Sorted	Having a systematic, logical route to decide what to do.
Psychic	'The aura is good'. 'This feels okay' style of decision-making.

⇨ TOP TIP

→ Reflection is better and reinforced when you write down your thoughts or speak them aloud.

2.2 Within module reflection

A module provides a useful focus for reflection. Developing your responses to **Try This 2.4** will help you to plan your study and provide some answers for job interview questions. Thinking about how you work in advance, and getting used to reflection (thinking, making notes to self …), gives you more chance to enthuse and be positive. You may not have had much experience of some skills, but any experience is better than none.

Try This 2.5 and **Try This 2.6** contain some more ideas. Different reflection styles suit different people and circumstances.

TRY THIS 2.4 – Using PDP reflection in modules

Read the module handbook and learning outcomes. List the skills (see **Try This 2.1**) you expect to use and develop. Pick one or two you want to focus on *actively* developing. (Doing the module should develop all the skills; this is about choosing specific areas to focus on, your choice). Create a brief action plan to achieve each goal. Get started. Make notes (reflect) about how you are developing every two weeks.

Draft answers to some of the following statements as mini case examples that you could use at an interview to support your statements. If you find this tricky, see some sample responses at the end of the chapter.

☺ What I want to get out of attending this module is …
☺ This module/session helped me to develop a clearer idea of my strengths and weaknesses, for example …
☺ I have discovered that when decisions are needed I …
☺ The skills I used well were … Other members of the group showed me new ways to …
☺ The preparation for my (the group) presentation was … My points to improve next time are …
☺ Our group could have done better if …
☺ I (our group) made decisions by … My role was …
☺ I have learned … about interview technique/asking questions/planning laboratory work/ investigating in the field.
☺ I most enjoyed … about the exercise/session/module/degree course.
☺ I least enjoyed … about the exercise/session/module/degree course. Next time I plan to …
☺ The biggest challenge to me in this exercise/session/module/degree was …

TRY THIS 2.5 – Reflecting on a day

Brainstorm a list of things that happened (five minutes maximum, a back-of-an-envelope list), e.g.:

> *Went shopping.*   *Haircut.*
> *Went to Dr Impossible's lecture.*   *Talked to Andy all afternoon.*

Then brainstorm a list of things that made the day unsatisfactory, e.g.:

> *Andy talked to me for hours.*   *Bus was late.*
> *I didn't understand what Dr Impossible was going on about.*
> *Printer queues were long.*

Leave the two lists on one side for a couple of hours. Then grab a cup of coffee, a pen, reread your lists and make a note about where you might have saved time, or done something differently. Consider what might make life more satisfactory if these situations happen again:

> *Natter to Andy for an hour MAXIMUM! over coffee and leave.*
> *Take a book on the bus.*
> *Have a look at Dr Impossible's last three lectures. If it still doesn't make sense I will ask my tutor or Dr Impossible.*
> *Need to take something to read, or do online searches, while printouts are chugging through.*

> ⚙ TRY THIS 2.6 – Reflecting on an activity
>
> These questions prompt reflection. Pick something from the last week and make some notes.
>
> ? What have I learned from doing this essay/presentation/report/mapping project …?
> ? How have my ideas developed in this essay/project/…?
> ? What worked least well? Why did it not go so well? What could I do next time to make this easier?
> ? How have I developed my ideas in this essay/project …?
> ? Was it worth the effort that I made? Did I have a good balance of reading/thinking/writing/revising/finalising?
> ? How has the feedback I got helped?
> ? How does this piece of work help me in preparing for future assignments and exams?
> ? Was the outcome worth the effort I made? Did I have a good balance of reading/thinking/writing/revising/finalising time?

All modules develop your skills. The course outline will include learning outcome statements such as: 'On completion of the module students will have …' Use these statements to capture examples and your involvement. They can be very straightforward:

- Time management skills: I completed …n… essays, and …y… projects always on time.
- Organisation: Organised a group project on …; I planned the mini deadlines that kept us on track to …

What examples do you have if asked about your:

Ability to put ideas across	Ability to work individually
Ability to prioritise tasks	Problem-solving experiences
Self-motivation	IT skills
Communication skills	Approach to networking

2.3 Prioritising – What to do first?

There are many competing demands on your time, and it is not always obvious whether the next research activity involves finishing a practical report, browsing Google Scholar or the library shelf for next week's essay, or reading another paper. Who or what takes most of your time? Some tasks take longer than others, but the proportions should be roughly right. Questions which encourage prioritising include:

? Why am I doing this now? Is it urgent?
? When and where do I work best? Am I taking advantage of times when my brain is in gear?

? Is the time allocated to a task matched by the reward? For example, it is worth considering whether a module essay worth 50 per cent deserves five times the time devoted to a GIS practical worth 10 per cent.

? How long have I spent on this internet search, seminar preparation, mapping practical, Africa essay? Were these times in proportion to marks available? Which elements deserve more time?

? Am I being interrupted? If I worked somewhere else would that help?

? Who causes me to take a 'time out'? How can I reduce unplanned stops by an hour a week?

Use **Try This 2.7** to test ideas.

TRY THIS **2.7 – Last week/next week?**

Considering last week, pick some of these reflection questions and consider where you might have acted differently. What else might you have done?
Thinking about tomorrow – how do these ideas influence your plan?
What about next week?

Figure 2.1 What students said about their university experience. (Wordle created using WordArt.com.)

Figure 2.1 shows a simplified Wordle created from the comments of final year students reflecting on their time at university. What are you looking for? How large or small do you expect some of these words will be in your reflections?

2.4 Reflection develops your employability – start now

Employers are looking for people with a mix of skills, evidence of your intellectual, operational, practical and interpersonal skills. Your intellectual skills are demonstrated by your degree certificate. You communicate your skills and attributes through your CV, letter of application and through self-advocacy at interviews (Chapter 6). Employers want enthusiastic, articulate and reflective graduates, those who can explain with examples, evaluating their experience and qualities. They seek people with the awareness and self-motivation to be proactive about their learning. The ability to teach oneself, to be aware of the need to update one's personal and professional expertise, is vital for effective company or organisational performance.

> *Filling in the portfolio just seemed a crazy waste of time at uni, but I just do it every day at work, it's so normal. It's how you keep track of everything you do.*

So much happens in life that it is easy to forget events and activities that have been developmental; however, effective job applicants talk fluently about them. Your PDP, e-portfolio, notes, memory box, diary, journal … (whatever you use) is the first place to look when applying for a job. Your record of your thoughts and actions from the **Try This** activities will remind you of what you did and the skills involved.

> *When we first talked about it our group agreed you just made it up and that is kind of what happens at school and here. You do it just for the teacher. Talking to Elle, you could see her line manager is the person she works with all day and talks to every day, so making it up isn't an option. And she was really positive about it helping her with doing a better job.*

Most large organisations have PDP processes as part of their staff development and appraisal. Keeping accurate records is critical for professional standards in a range of jobs. Earth scientists, technologists and water engineers have professional body e-portfolios to complete throughout their careers. Where possible, university portfolios mirror your professional body portfolio (CIWEM 2018, Geological Society 2018).

> *The session reminded me to start thinking about getting some work experience this summer, that I need to do something now.*

Whether you are expert in modelling the spread of disease, have researched Mongolian housing patterns or abseiled down a glacier is of little interest to most employers. What is relevant is that faced with the task of researching the market for a new type of chocolate, you can apply the associated skills and experience gained through researching the nineteenth-century development of the Co-operative Movement in Rochdale, or new waste control management processes (thinking, reading, researching, presentation, making connections, reflecting) to designing and marketing cocoa products. It is your ability to **apply** the skills acquired through school and university **in a workplace role** that employers value.

Reflection can highlight the absence of a particular 'skill'. There is time to get involved in something that will demonstrate you possess that skill before the end of your degree. When evaluating your skills include those acquired through leisure pursuits or work experience. Driving, shorthand, stocktaking, flying, language skills, writing for a newspaper or magazine, treasurer, secretary and chair of societies ...; they all involve skills – time management, negotiation, listening, writing reports and many more. Work experience does not have to be paid work; voluntary activities provide valuable experience that makes you very employable (Chapter 6).

2.5 References and resources

Search: *PDP, personal development planning, reflection, career planning, lifelong learning, graduate skills, volunteering, internships, career development.*

Bassot B 2017 *The Reflective Journal*, (2nd Edn.), Palgrave, London.

CIWEM 2018 Continuing Professional Development, Chartered Institution of Water and Environmental Management, http://www.ciwem.org/membership/applications/ Accessed 15 December 2018.

Geological Society 2018 Continuing Professional Development (CPD) & Training, www.geolsoc.org.uk/Membership/members/CPD-and-Training Accessed 15 January 2019.

Moon J 2006 *Learning Journals: A Handbook for Academics, Students and Professional Development*, (2nd Edn.), Routledge, London.

PebblePad 2018 Introduction, http://www.pebblepad.co.uk Accessed 15 January 2019.

University of Bristol 2018 Personal development planning (PDP), https://www.bristol.ac.uk/students/study/pdp/ Accessed 15 January 2019.

University of Plymouth 2018 Personal development planning (PDP), https://www.plymouth.ac.uk/your-university/teaching-and-learning/guidance-and-resources/personal-development-planning-pdp Accessed 15 January 2019.

University of Sheffield 2018 Personal development planning, https://www.sheffield.ac.uk/lets/toolkit/support/dev-progs Accessed 15 January 2019.

Quick crossword 1

Each of the answers are very loosely related to reflection. Answers in Chapter 28.

Across

1 Previously
4 Instruction, guidance
6 Capacity, fitness
8 Echo, reflect

Down

1 Book cover publicity
2 In good shape
3 Very long time
5 External
6 Purpose, target
7 Dickensian exclamation, left or right?

Answers: Try This 2.4 – Reflecting on a class or module

Reflections from various second-year students:

What I want to get out of attending this module is … *I want to realise my full communication and research skills / I would like the module to give me a clearer insight into a topic I enjoy and think I might want to pursue as a career / I want to improve my computing skills.*

I have discovered that when decisions are needed I … *I now realise that I make more decisions than I realise, but in general I try to avoid the process if possible / I am not particularly decisive, but when I have to make a decision I think things through very carefully / I adopt different decision making processes in different cases / I know it is a very weak point / I think I will find it helpful to understand more about how I make decisions in order to improve. I probably agonise too much / Sometimes I am really rational and think things through, other times I am totally impulsive.*

The skills the group used well were … *As a group we had good discussions, no one person took on the role of leader. Everyone listened to everyone else and everyone had valid points to make / I took part in the discussion, however some members of the group did this better.*

The preparation for the group presentation was ... *To start with we had a distinct lack of preparation, hadn't read the briefing material, and generally didn't know where to start. We overestimated the degree of detail required which explains why we took so long. Organising what we were supposed to do and deciding who should do what, wasted some time.*

Our group made decisions by ... *general agreement, but mostly one person took the lead and we all just went along with it. We had some texts that got us confused and no one was doing much until the last week. We didn't look at ... (VLE page) which we realised at the end would have helped. Looking back my three words to describe us are chaotic, muddled, unfocused – and we got 64% – result!*

What did you enjoy least about the exercise/session/module/degree course? *There was a lot of information to evaluate in a short time, a bit of quick thinking required. Some aspects were boring and appeared unnecessary. Managing time was difficult. / The lectures were OK but the pracs seemed really tedious. There was loads to do ... When we got to the project work, then the lecturer assumed we could do the analysis because we had had the pracs. It was obviously really useful then. ... Lecturers tell you stuff will help with dissertations and projects but you don't really believe them until you have to use the stuff later.*

3

Balancing study, work and fun

I plan to be spontaneous tomorrow!

University is different from school and work. There is lots of free time for running, acting, being elected union president, playing the lute, ballroom dancing and chatting. How are you managing – are you organised or a procrastinator? Watch Tim Urban's (2016) TED Talk on procrastination to find out. This chapter has ideas to help you to meet deadlines, have fun and avoid panicking at the last minute.

Time-management techniques are especially vital for people with time-consuming sports training, part-time jobs and family commitments. Developing your time-management skills should allow you to do all the boring tasks like laundry and essays, leaving time for other activities. It is unlikely that any one idea will change your life overnight, but a few time-saving short cuts can relieve the pressure. Try something. Use your reflection and evaluation skills to identify what to do next and to assign time to research and reading.

Think of this as **project management**. Your life is your big project; an essay, module, field project and mapping exercise are small projects. Ideally one envisages the research/thinking/reading for an essay, project or dissertation moving linearly from inception to final report or presentation. Regretfully the process is rarely this simple. The normal elements of life intervene, and the way you understand a topic changes as the research progresses. You need plenty of time for the research/thinking/reading process to evolve. Halfway through your research you may have to go back almost to the start, reconsider your approach and start a revised programme (Figure 3.1). So, plotting and planning your time, leaving lots of space to recognise and adjust to changing goalposts, and fit in the unexpected, is a great university and life skill. There are many apps for planning, but first find out the approach that suits you. Start with **Try This 3.1**.

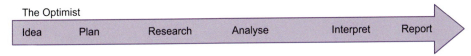

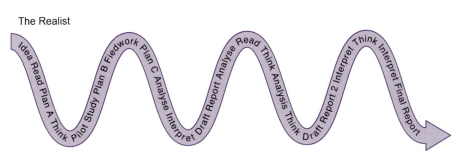

Figure 3.1 The practice of learning

TRY THIS 3.1 – Project manage your tasks

Find some Post-it notes, ideally in five different colours for: the Deadline date; Tasks; Milestone targets with dates; Risks; and People who can help.

Now plan a holiday visit to … on **date** … Put all the **tasks** that need doing on one colour of Post-it (research, book, ask friends to come), use another colour for **milestone targets** (visa obtained, booking made), and another for **risks** that may disrupt your plans (no money, airline closes, cannot get time off work). The risks show where more planning will help (more task notes) or a plan adjustment (move things around) will be important. Add in names of **people** who can help you on another colour.

Put the Post-its on a large piece of paper/the wall/door … and move them around to get them into order as a flow diagram that shows what needs to be done, and when. Involve other people – can they improve your plan? Other people make it fun and will prompt additional ideas. The advantage of Post-it notes is that you can move them around as you think of additional tasks, or deadlines are missed (this happens). And you feel good when parts of the project are completed as you progress.

Then use the same technique to plan your next essay or project.

See Figure 3.2 for example charts.

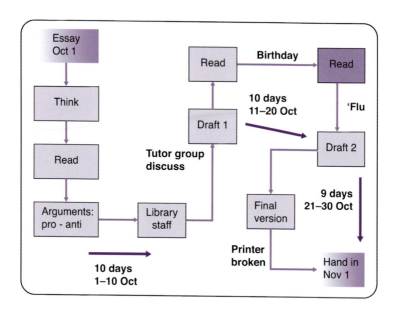

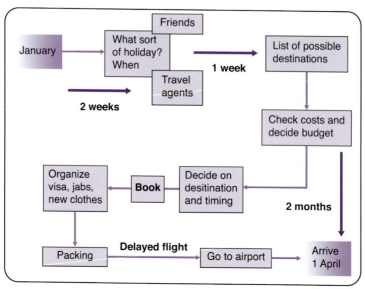

Figure 3.2 Project charts to complete an essay, and book a holiday

3.1 Is there a spare minute?

There are many templates for planning, but you have to start somewhere. Use **Try This 3.2** as a start. Assuming that social and sport activities will fill every night and all weekend, and that arriving at university before 10.00 is impossible, the remaining time is available for research, reading, thinking, planning and writing, without touching the weekend or evenings. If you add a couple of evening study sessions to the plan it will save money, due to temporary absence from bar or club, and help to get essays written. Divide this total free – oops, I mean *study* – time by the number of modules to get a rough target of the hours available for working on each module.

TRY THIS 3.2 – What spare research and study time?

Fill in your timetable: lectures, practicals … the works. Block out an hour for lunch and a couple of 30-minute coffee breaks each day. Add up the free hours between 10 and 5 to find your Total Learning Time.

	Morning		Afternoon		Evening
Monday					
Tuesday					
Wednesday					
Thursday					
Friday					
Saturday					
Sunday					

3.2 What do I do now?

Confused? You will be.

There are lots of planning techniques – you need to find what works for you.

Diaries and timetables

University timetables can be complex, with classes in different places from week to week. A diary is essential. A weekly skeleton timetable will locate blocks of time for study (**Try This 3.2**). Use it to allocate longer free sessions for tasks that take more concentration, like writing, reading and preparing for a tutorial or workshop. Use shorter, one-hour sessions to do quick jobs, like tidying files, sorting lecture notes, summarising the main points from a lecture, reading a paper you downloaded/photocopied earlier, highlighting urgent reading, doing online searches, thinking through an issue and making a list of points that you need to be clearer about. Don't be tempted to timetable every hour. Leave time for catching up when plans have slipped.

Lists

Sort out what you need to do under four headings: Urgent Now, Urgent Next Week, Every Week and Fun. If you tackle part of the non-urgent task list each week, you will be less overwhelmed by Urgent Now tasks at a later date. Have a go at **Try This 3.3**.

Plot your activity for the next three weeks under these headings. Put a * against the items that you want to do in the next four days, and make a plan.

Urgent now	Urgent next week	Every week	Fun
Wednesday, Essay Urban Poverty Friday, Paleontology Report 3	Read for tutorial on Nutrient cycling Find out about wind farms	Tai Chi Ironing Supper	Friday, cinema Thursday, Dan's party, get card and beers

Create a diary template, online, on your phone or tablet, or in your workbook with your regular commitments: lectures, tutorials, sport sessions, club and society meetings ... This provides the skeleton for planning. If weekly planning is too tedious, go for the 30-second breakfast-time, back-of-an-envelope version. It can really assist on chaotic days when classes are spread across the day, which encourages the time between to disappear. There are free hours but 'no time to do anything properly'. Completing short jobs will avoid breaking up days when there is more time. Try to set your day out something like this:

9 Lecture	10 Coffee Annie and Dan	11 Lab, Computer practical	12 Finish computer practical, Lunch	2 Tutorial	3 Sort file, Read last week's Africa notes	4 Africa seminar	5–9 Shop, night out

On days with fewer classes, two or more free hours give you good research time for concentrated study. A day might look like this:

10–12 Read and make notes for Marine Pollution essay	12 Lecture	1 Lunch, email	2 Computer practical	3–5 Read and make notes for Marine Pollution essay	5–9 Film, Phone calls

Knowing what you want to do in your research/thinking/reading time, saves time. Deciding at breakfast to go to the library after a lecture should ensure you go to the right floor with your notes and reading lists. Otherwise you emerge from a lecture, take ten minutes to decide you would rather read about ecotourism than sustainability, discover you haven't got the ecotourism reading list, then look at the sustainability list to decide which library and floor to visit. All this takes 45 minutes and it is time for the next class.

3.3 Tracking deadlines

Deadlines are easily forgotten. Some people carry a term or semester chart that highlights deadlines. Figure 3.3 shows two styles. The first is essentially a list. The second shows where pressure points build up; in this example, Weeks 10 and 11 already look full. The computer report due on Friday needs finishing before the Ball on Thursday! This second style highlights weeks where personal research/thinking/reading time is limited by other commitments. Start planning backwards. How would you adjust the tasks and timing in Figure 3.3 to meet the deadlines while having the worst flu ever and losing your USB stick with all the notes? The Post-it note wall-planner helps you here because you move the notes around, rather than rewriting the plan.

Module	Assessment	Hand-in Date	Interim deadlines
EOE1202 Life, Energy and Matter	Essay	10 Dec	Check out a couple of background texts and case studies by 25 Nov First draft by 1 Dec Think, create diagrams and revise 6 Dec Final draft 8 Dec
GEOG1070 Statistics	Practical	16 Nov	Sort out the data set and run Nov 1 Draft report Nov 5 Final report Nov 10

Week No/ Date	Fun	Workshops	Computing	Tutorial	Essays	Laboratory
2/ 6 Oct			Report 1 Friday			
3/ 13 Oct						Climate report Wednesday
10/ 1 Dec	Rock Soc Ball Thursday	EOEE1040 Worksheet Tuesday	Report 6 Friday	Presentation Aquaculture Tuesday		
11/ 8 Dec	End of term Christmas shopping	GEOG1060 Test Tuesday	Report 7 Friday		National Parks essay Thursday	Sediment prac. report Thursday

Figure 3.3 Example deadline planners for the semester

3.4 What next?

The sooner you fall behind the more time you'll have to catch up.

Keep reviewing your plan. Look at the relative importance of different activities, so that you don't miss deadlines. Have a go at **Try This 3.4** to practise prioritising. Reflect on where you could re-jig things to release two lots of 20 minutes. Twenty minutes may not seem much, but using this time to sort notes, reread last week's lecture notes or skim an article, is 20 minutes more than you would have had.

TRY THIS 3.4 – Priorities?

Think about yesterday: jot down the hours or minutes different tasks took. Change the list to suit your activities. Considering the priority each should have had, score each 1–5 (1 = high priority, 5 = low priority). How do your priorities match the time taken? This is an activity where you are **feeding back** to yourself on your activity.

Life	Hours	Priority	Degree	Hours	Priority
Cooking and eating			Talking about ideas		
Sleeping			Reading		
Shower/Dressing			Browsing in the library		
Exercise			Thinking		
Travelling			Sorting lecture notes		
TV			Writing		
Reading for fun			Lecture attendance		
Cleaning the flat			Computer practical		
Washing/Ironing			Laboratory practical		
Phone calls			Planning time		
Social media			Online searches		

Tackling **Try This 3.4** might encourage you to use a day planner in Figure 3.4 style. Write a 'to do' list for tomorrow, then prioritise your activities. What is important? What should be done first? Can you double-task, reading while the washing tumbles around? Put some times against the activities. Ticking off completed jobs feels good!

Priority	MUST DO!	When
1	Post Mother's Day card	On way to college
2 =	Read Chapter 3 – Wave Power by Sue Nami	10–12
2 =	Launderette	10–12
3	Look at ecology notes for seminar (Rm 6.13)	12–1
4	Check email	After lunch
1	Lecture 2.00 (Lecture Theatre 4)	2.00
5	Spell and grammar check tutorial essay	After supper
6	Sort out practical notes	After supper
7	Make a list of jobs for tomorrow	After supper

Figure 3.4 An organiser like this?

If you can do a task immediately and easily, **Just Do It**. Generally it helps to allocate larger tasks to longer time chunks and leave little tasks for days that are broken up. Do not procrastinate: '*I cannot write this essay till I have read …*' is a lousy excuse. No one can ever read all the literature on any topic, so set a reading time limit, write, and then go clubbing.

Do you do:

1 Hard tasks first?
2 Easy tasks first to warm up to tricky ones?
3 Make a list and then do tasks at random?

Do what suits you, there is no right way. Find out what works for you.

TOP TIPS

☺ Get a timetable/diary/set up the alarm or timer on your phone, tablet or laptop.
☺ Plan weekends and time off well ahead; sport, socialising and shopping are critical. Having worked hard all week, you need and deserve time off. Following a distracting, socially rich week, maybe there is time for some study. Sunday can be a great time for reading, thinking and drafting because very few people interrupt.
☺ Filing: '*… so many modules, so many handouts, my room looks like a recycling depot*'. Take ten minutes each week to sort out notes and papers in your room and computer.
☺ Decide when and where you work best: your high and low periods of concentration. High concentration times are best for study. Low

concentration times are great for washing dishes. In noisy spaces do easy tasks like planning. If your mates are chatting in the geology library, go to the biophysics library. Avoid interruptions in your high concentration time.

☺ What do you find difficult? Sort out these tasks, do them first in your high concentration time. Divide awkward tasks into manageable chunks and tackle each one separately. Finishing parts of a task ahead of time gives you time to think about interpretations (better marks). Most tasks that seem difficult become more difficult because they are left until time is short, and time pressures make them more tricky to do well. Get them out of the way.

☺ Short study times are good. Break up your day into one- and two-hour blocks. Set the timer on your phone or laptop, work hard at reading and writing, then have a break.

> *Sorting out my notes each Sunday made a fantastic difference to year 2. I could find stuff.*

☺ Double-task – view 'dead time', when walking to university, doing the laundry or cleaning the bathroom, as 'thinking opportunities'. Listen to a podcast, plan an essay, mentally review lecture ideas …

☺ Vacations. Recover from term. Have a really good holiday. Think about dissertation possibilities.

☺ Time has a habit of drifting away. Can you limit lost time when the pressure is on? Minimise walking across campus. Ask 'Is this a trip I need to make? Could I be more efficient?' Make an agreement with a friend to do something in a certain time and reward yourselves **afterwards**.

☺ Be realistic. Most days do not map out as planned, things (people) happen, but plans make you more effective most of the time.

The 24/7/365 trap

When lecturers mention 'blended learning', they mean that you will have a blend of face-to-face, traditional reading and study, and online activities. It is extremely helpful that your learning resources are available 24/7 online, and the library may be open day and night every day of the year, but this ubiquitous availability can pose a problem. You can study at any time, potentially making you less organised than students who can only access the library Monday to Friday, 9 to 5. Sorting out regular times for study is critical. Online learning support lets you explore topics in more detail and total privacy. Blended learning enables you to manage and personalise your learning, so your university experience is unique, but you also need sleep!

Try some of these ideas, give them a real go for three weeks. Then reread this chapter, consider what helped and what did not, and try something else. Find a routine that suits you and recognise that a routine adopted in your first year will evolve. A realistic study timetable has a good balance of social and fitness activities. Don't be too ambitious. If there was no time to read last week, finding 30 minutes to read one article this week is a step forward. If you want an online planner try the calendar function on your computer and phone, and there are many time-management apps. Search for 'time management for students' or 'student planners' and look critically at the reviews. Look for something that is free and easy to use. **Due** and **Studious** are two of many apps that will help you track deadlines, set reminders and set off alarms to prompt action, but the tech can get in the way of real planning and action. Be careful that you are not being satisfied by having set up the app and entered data, but not actually acted on the prompts!

JUST DO IT

3.5 References and resources

Allen D 2015 The art of getting things done, http://gettingthingsdone.com/ podcasts/ Accessed 16 January 2019.

UNSW 2010 Support with time management, The Learning Centre, The University of New South Wales, https://student.unsw.edu.au/support-time-management Accessed 16 January 2019.

Urban T 2016 Inside the mind of a master procrastinator, https://www.ted.com/ playlists/534/how_your_brain_functions_in_different_situations Accessed 16 January 2019.

Vanderkam L 2016 How to gain control of your free time, http://www.ted.com/ talks/laura_vanderkam_how_to_gain_control_of_your_free_time Accessed 16 January 2019.

Drop out 1

Remove one letter from each column to make one seven-letter word, the remaining letters will align to make six seven-letter words across. Answers in Chapter 28.

M	A	X	M	O	T	H
M	E	M	M	I	A	N
P	U	R	I	R	E	E
G	E	M	T	E	A	U
P	L	A	T	O	P	N
I	S	O	T	E	A	E
G	E	O	D	C	S	Y

4

Looking after yourself, commuting, mature study and well-being advice

Crafting the life you want takes thought, time and insights from others – staying healthy, exercise, eating vegetables and sleeping are critical.

This chapter is relevant for all students at all times. It's important to look after yourself, and to notice, and maybe help, when others are struggling. University should be exciting and challenging, but there may be times when it presents difficulties. Being away from family and friends, in a big community of students, living on your own, is challenging. Commuting and mature students also fit in travel and other commitments. There are so many different things you could do, it's confusing. Practising reflection (Chapter 2), and time management (Chapter 3), will help to identify and manage your concerns, and particularly to identify opportunities to seek support. If in doubt, ask! Whatever they are called on your campus, the staff in the Wellbeing Centre/Student Services team/Chaplaincy/Health Centre come to work every day to help students have the best possible time at university. Being healthy and well enables you to study effectively. If you are feeling unwell, stressed or unable to cope, please contact your student services.

It's vital to remember that higher education is adult education, where the staff respect your right to make personal decisions. University staff will provide lots of support, advising (not instructing) you about what to do when you ask. You are in control of your experience. So if (when) you are confused, concerned or worried, who do you ask? Use **Try This 4.1**.

Advisers are generally very, very experienced in working with students and are aware of the services available in the local area through general practitioners, hospitals and voluntary organisations. Support options you may be offered include conversations

---

TRY THIS 4.1 – My university support

Find out what your university provides, and where to find people. Typical terms to use in web searches include: student support, well-being, health, happiness, counselling, learning support, learning development, disability service, welfare support, health and wellbeing centre.

Wander in and introduce yourself. Find out about the workshops, resources and internet support they offer. Advice on writing, revising, exams … cooking and travel is hugely useful for everyone.

face-to-face or by phone, counselling, coaching, workshops, small-group discussions, videos, podcasts, and information online, in books and leaflets. Meanwhile, here are my top tips for a happy, healthy, productive, enjoyable and rewarding experience.

⇨ TOP TIPS

For enjoying university study:

☺ Talk to everyone: people sitting next to you in class, on the bus, in halls, cafés and … Sociable relationships take you out of your bubble and enable you to appreciate others' points of view.

☺ Take/make time to read and think. Slow down! There is no need to rush everything. As with slow, mindful eating – take time to appreciate and enjoy your learning. Turn off the TV and anything else that is distracting, be mindful, concentrate on what you are doing and appreciate the new stuff. Ask (and make a note of) 'what I know now that I didn't know an hour/day/week ago?'

☺ Fresh air, walking, some exercise (most days) refreshes your brain; this helps thinking and will help you to sleep. Yoga, Tai Chi, basketball, rugby, dance … will be fun and help you keep well.

☺ Start a new activity, sport or hobby. Joining walking/skiing/climbing/surfing groups will take you into the fresh air. Whatever you choose, it's good to get away from studying one day a week.

☺ Use reflection techniques and planners (Chapters 2 and 3) to analyse how much time and attention you give to social media. Is screen-time interfering with your opportunities for enjoying university?

☺ Fruit, vegetables, and balanced meals help your energy levels, thinking processes and mood. Eating regularly is important.

☺ Sleeping well allows your body to recover and your brain to organise ideas and information. We know that losing sleep affects your learning. Aim for a balance of late nights!

4.1 Putting 'learning' support in place

Successfully obtaining a university place is brilliant, congratulations. If you have any sort of physical, learning or welfare issues it is important to seek advice from university services before arrival, or as soon after as is possible. Please bring your health records with you.

These records provide the evidence for the university to make 'reasonable adjustments' to your teaching and assessments. Where you have a recognised condition or diagnosis, the evidence of the support you were given in the past enables support to be organised relatively quickly. In large universities, where there can be many thousands of new students requiring support each year, getting involved early will be a bonus.

4.2 Moving to another country

Moving to a new country to study for your degree, or for a year as an international experience, can be both exciting and overpowering. The anxiety you experience is termed 'culture shock'. It's a normal process because everything is new – food, language, transport, weather, people and clothing. You can expect to feel confused, sad, lonely, have a temporary drop in confidence, difficulty sleeping and be moody. Remember you had the confidence to apply to study in another country. That was a brave, ambitious and confident decision. You can cope, it just takes time to adjust to so many new things.

The reflection activities (Chapter 2) can help you understand what's going on. Using social media, text and Skype to keep in touch with family and friends is vital. Eating healthily and getting regular exercise will help you be fit and enable you to meet people and build your network of friends. On campus the Chaplaincy or Faith Centre will connect you to faith activities of your choice. There is plenty of help, but you need to ask for it. Start at your Wellbeing Centre or Chaplaincy.

UKCISA 2018 Changing lives for good, UK Council for International Student Affairs, https://www.ukcisa.org.uk/ Accessed 18 January 2019.

University of York 2018 International student support, https://www.york.ac.uk/study/international/support/ Accessed 18 January 2019.

4.3 Homesickness

Anyone can experience homesickness, and many students do. You don't have to be far from home to feel anxious and stressed. Everything is new, your normal life activities are disrupted, and you're making lots of decisions. This is tiring, so sleep is very helpful. Homesickness is most usually experienced in the first couple of weeks of the academic year, and after the first vacation (Christmas holidays for northern hemisphere universities). Typical symptoms include feeling lonely, isolated, nervous, angry or sad. Headaches, lack of appetite and lack of concentration and the feeling that 'it's all too difficult', combined with loss of sleep can all make you feel insecure and anxious.

Remember this is normal. You knew university would be different (daunting, challenging, tricky); it's time to get involved. Smile. Big Smile ☺. Joining clubs and societies helps you meet people. Talk to people at induction, freshers fair, cafés, bars … Everyone else is nervous too. Get out of your room as much as possible, take coursework into the library to work with other people around.

Follow up **Try This 4.1** by visiting the various services: the library and careers, volunteer, union, international and … centres. They are full of people happy to listen and to cheer you up. Don't forget to call or message friends and family. Homesickness usually disappears in a week or so, as you adjust to your new way of life.

4.4 Studying from home

You are not alone! The number of students choosing to live at home while studying has changed significantly as higher education has become more expensive, and many mature students study from home. In some countries living at home is the usual circumstance; in the UK the numbers are over 40 per cent in some universities. Living at home is not 'second best'. It's usually a very good option financially, and friends and family are available to provide support, which can be exceptionally helpful. It just needs a bit of thought and planning to make sure you integrate effectively into higher education, whether your travel takes the same time as it did to school, or more than two hours each way. *'I didn't really understand what the university was offering, because I got the same bus as I did going to school, just got off a few stops later. There was a timetable, so I knew when to go to class, but you didn't have to be in the study centre, and no one was checking up like at school'*.

If you are one of the large group of students who commute up to a couple of hours a day each way, some organisation, time management (Chapter 3) and enthusiasm will be beneficial, and there are advantages. *'Living at home enabled me to continue with my part-time work throughout the year'*; *'… two hours on the train each way took quite a lot of getting used to, but I could always get a seat, my laptop fits the drop-down table, there's Wi-Fi, and I do all my reading there'*.

You have time to think through and plan your university activities, because you don't have all the hassle of moving to a new place, finding somewhere to live, moving into halls or flats, and working out where to shop and how to cook. While you are commuting, you can be getting ahead with reading and thinking, while campus-based students are still in bed, cooking, shopping … Your learning can be ahead of the class. Use **Try This 4.2 now!**

> **TRY THIS 4.2 – Commuter student support**
>
> What is available on your campus? Website search terms vary, try: *commuter, local, living at home, home-based, off campus, stay education, study at home* and *day student support.* Then search for: *learning support, learning development* and *library service* to find the support workshops, courses and software offered to all students. **THEN USE THEM!!**

Higher education is about more than attending classes, improving your skills and knowledge and performing well in assessments. As in all parts of life, feeling comfortable and engaged is important and if you only attend class and go home, you're likely to feel isolated, uncomfortable and have a sense of 'not fitting in'. University staff understand that to be successful in your degree studies, you need a sense of belonging. Staff arrange social and other activities to encourage networking, development of friendships and study groups. Perhaps inevitably students who study from home participate less in extracurricular activities, but these do have the benefit of providing skills development and enable you to develop your confidence in new or unfamiliar social and workplace spaces. Choosing to ignore all the 'extra' university opportunities may feel 'safer' and cheaper, but potentially adds to your isolation, and not getting as good results. *'In the first week there was this tour around town, I didn't go because I've lived here all my life. I didn't go to the freshers fair because I was carrying on playing basketball with the (local) team, so I wasn't interested in university sports. In the first lecture lots of people were talking to each other already, and I didn't … I guess I'd missed out on those, umm getting to know people … things'.*

On the other hand, *'I've had a crazy time at uni. Living at home with mum meant I had loads of time to get involved with climbing and squash and volunteering. No shopping or cooking, met loads of people, no hassle with landlords … Cool'.*

Sort your travel costs early

Look at the university website for advice on travel, including discounts on buses and trains, and check out the bus and railway websites. A monthly or term-time pass is probably more economical than buying individual tickets.

University study is more than class attendance

There is a temptation to feel that you only need to be on campus when you have a class. Each module generally involves ten hours reading and thinking for each class-based hour. Campus time gives you access to the library and computer laboratories. These are spaces for study, discussion, seeing what other people in your course are reading, and personal study. Communal learning is important. A day with just one timetabled class can seem like a waste of time – and expensive – but it offers uninterrupted library time, group work time, and with your monthly travel pass you have paid for your travel anyway.

Get the software to work at home

Visit the library and learning support website for the workshops, videos and handouts which explain how to use your university software, library, webinars … really efficiently. These enable you to work and learn flexibly on and off campus. If lectures are recorded make sure you can download and access them from home. Each module will have a resources site. Travelling time can be good for reviewing and revising lecture materials.

Make sure your tutor knows about your circumstances. If there are multiple laboratory classes for example, sign up or ask to change to the session which best suits your travel. Most staff have 'business hours' or 'drop-in' times when any student can seek advice. If these times don't suit your travel arrangements ask if you can Skype or phone instead.

Finding fellow commuters

Some universities have introduction and orientation events to help commuting students settle in and meet people. There may be a common room with lockers and a kettle especially for you. Check out social media and your Students Union for advice and events. If you have a long commute you will begin to recognise people. '*There are four of us who travel together, all different degrees ..., and we talk about each other's courses and essays which is really interesting*'; '*The guy in the year ahead of me was really helpful about getting cheaper tickets, and what to expect ... Meeting him made me feel much more okay with it*'.

> ⇨ TOP TIPS
>
> ✓ A Thermos flask or insulated mug for a warm drink and interesting sandwiches make journeys more pleasant. Keep hydrated.
> ✓ If your bus or train is noisy – earplugs!
> ✓ Back up your computer every week, and save copies of your assignments to the cloud and USB sticks as you go. Losing your laptop is bad enough without losing your work too.
> ✓ Plan your work times and time off, and make time to have coffee with other students on campus.
> ✓ Travel light ... Carrying unnecessary stuff is exhausting. Laptop or notebook, do you need both?
> ✓ For long journeys a mix of eBook, textbook and fun reading will cheer you up.
> ✓ Plan time off at weekends, travelling is exhausting.

NUS 2015 Reaching home: policy and practice for students living in the parental home, https://www.nus.org.uk/PageFiles/12238/Reaching%20Home.pdf Accessed 16 January 2019.

The Student Room 2018 Tips for commuting students, https://www.thestudentroom.co.uk/university/life/tips-for-commuting-students Accessed 16 January 2019.

University of Birmingham 2018 Local or commuting students, https://www.birmingham.ac.uk/welcome/welcome-week-and-beyond/accessing-support/localorcommutingstudents.aspx Accessed 16 January 2019.

University of York 2018 Local and commuting students, https://www.york.ac.uk/students/support/off-campus-students/ Accessed 16 January 2019.

4.5 Mature students

(Not to be read by anyone under 21 – Thank you)

Worries

Consider the student comments and responses here. What is your position? What actions might you take?

'*They are all so young, and so bright and I don't think I can do this*'. Oh yes you can. You have had the bottle to get your act together and make massive arrangements for family and work: handling a class of bright-faced 19-year-olds is easy. Keep remembering that all those smart teenagers have developed loads of bad study habits at school, are out partying most nights, are fantasising about the bloke or girl they met last night/want to meet, and haven't got your incentives to succeed. Your degree is taking time from other activities, reducing the family income and pension contributions; great motivations for success. Tutors know most mature students work harder than students straight from school, worry hugely and do very well.

'*I really didn't want to say anything in tutorial, I thought they would laugh when I got it wrong*'. I'm nervous, you are nervous, he … (conjugate to gain confidence or get to sleep). At the start of a course, everyone in the group is nervous. Your experience of talking with people at work, home, office, family, scout group … means YOU CAN DO THIS. It is likely that your age gets you unexpected kudos; younger students equate age with experience and are likely to listen to and value your input.

'*I hated the first week, all those 18-year-olds partying and I was trying to register and pick up the kids*'. The first term is stressful, but having made compromises to get to university, give it a real go. If life is really dire, you may be able to switch to part-time for the first year, gaining time to get used to study and the complexities of coping with home and friends.

'*I seem to be running our practical group work?*' Mature students end up leading group work more often than the average. Your fellow students will be very, very, very, very happy to let you lead each time as it means they can work less hard. Make sure they support you too!

'*It takes me ages to learn things, my brain is really slow*'. OK, recognise that it does take longer for older brains to absorb new ideas and concepts. The trick is to be organised and be ACTIVE in studying (Chapter 1.4). This whole book should prompt useful ideas. The following suggestions are gleaned from a selection of mature students. It is a case of finding the tips and routines that work for you. Like playing the cello, keep practising.

Remind yourself why you decided to do this degree, then ask: '*is this statistics or computer practical really worse than anything else I have ever had to do? OK, so it is worse, but it will be over in X weeks*'.

TOP TIPS

☺ Review notes the day after a lecture, and at the weekend.

☺ Check your notes against the texts or papers; ask 'do I understand this point?'

☺ Practise writing regularly – anything from a summary paragraph to an essay. Write short paragraphs, which summarise the main points from a lecture or reading, for use in revision. Devising quizzes is fun (Chapter 24, Try This 24.3–6).

☺ Talk to people about your study topics – your friends, partner, children, people you meet at bus stops, the hamster. It's great practice in summarising material, and explaining in an interesting manner raises your interest levels. People often ask useful questions.

☺ Visit the library, online or in person on a regular basis: timetable part of a day or evening each week.

☺ Meet a fellow mature student once a week or fortnight to chat about experiences and coursework over lunch, coffee or a drink.

☺ Have a study timetable, and a regular place to study – the shed, attic, bedroom or a corner of the hall. Once the family knows you are out of sight, they will find devious things to do unsupervised. Stick a notice on the door:

KEEP OUT

MUM's WORKING

☺ Make specific family fun time and stick to it.

Some time management and a bit of organisation will get you through the first year, and by the second year you will know so much more about the subjects and about how to manage. When you have finished and got your degree, your family will be amazed, stunned and you will have the qualifications you want. Don't feel guilty; lifelong learning is the way the world is going, and there is increasing provision and awareness of mature students' needs.

As a family we really tried to go out once a week for an hour, an evening, or all day. This made a real difference. We did walks, swimming, supper at the pub, visits to friends and family and loads of odd things. We took turns to choose where we went, which the kids thought was great.

Support

Universities have services which support all students. These have various names (Learning Support; Additional Learning Support; Student Health Care; Student Advice; Wellbeing Centre; Student Learning Advice Team, Writing Support ...), and staff who will be delighted to see you. Their online and face-to-face support can save you lots of time worrying about how to make university work for you. Good people for a quick chat.

In your department, key people are your personal tutor for academic and pastoral support, the administrators in the School office, and your fellow students of all ages.

Many mature students think the Students Union '*is for younger people.*' BUT the Union often has cheaper food, coffee, a shop that can be very competitively priced and groups that might be called the Mature Student's Society, or Mature Student's Network. Check it out.

Mature Student Mentoring Schemes run in some universities, where students who have already survived the first year offer support. Just meeting for a cup of tea and a chat is reassuring and a chance to get answers to questions you don't want to ask your tutor. You may feel that alumni–student mentoring is really for younger students, but the opportunity to talk with people from other businesses, develop your networking skills and personal network of contacts may be very helpful in deciding what to do after graduation.

Problems?

Check out your Students Union. Most Union welfare offices have advice sheets for mature students, advisors for mature students and a mature students society. If there is no Mature Students Society, start one. Generally, academic

> *I was pretty nervous, but the lecturers were friendly and pleased to see you.*

staff are very experienced in student problems and how to solve them. Most departments have a staff tutor or contact for mature students. Go and bother this nice person sooner rather than later. If you feel really guilty, buy him or her a drink/chocolates later!

Families

Even the most helpful family members have other things in their heads besides remembering that they promised to clean out the rabbit cage, vacuum the bedroom or buy lemons on the way home. On balance, a list of chores that everyone agrees to and adheres to on 40 per cent of occasions, is good going. Thank the kids regularly for getting jobs done, and remember your degree should not take over all their lives. YOU NEED to take time away from study too, so plan in trips to the cinema, games of badminton, visits to sports centres, and get away from academic activities.

Suspension or short courses

This should be the last resort. Talk to your tutors: university staff are very experienced in offering support and can suggest options to handle difficulties well in advance. It is often possible to suspend studying for a semester or a year, or to stop after one or two years. Most universities offer some certification for the successful completion of each whole year. (See section 4.12 below)

Dawson C 2006 *The Mature Student's Study Guide*, (2nd Edn.), How to Books, Oxford.

Khambatta C 2013 *Mature Students: The Essential Guide*, Need2Know, London.

Open University 2011 Advice for mature students, https://help.open.ac.uk/advice-for-mature-graduates Accessed 16 January 2019.

Sheffield Hallam Students Union 2019 Mature Students, https://www.hallamstudentsunion.com/advice_help/maturestudents/ Accessed 16 January 2019.

4.6 Faith and spiritual support

Universities around the world have multi-faith chaplaincy services which provide support for students of all faiths and none. Typical campus Chaplaincy Centres have prayer spaces, meeting space, comfortable seating and ready supplies of tea. Some of the chaplains have roles in local churches and mosques, others are academics who volunteer with the chaplaincy team so all faiths are linked up.

Importantly, you don't have to be religious to get involved with chaplaincy activities. Chaplaincies are characterised by their warm welcome for everyone and capacity for listening. If you need to talk, the chaplaincy is a great place to start. Chaplains offer pastoral support which is confidential and non-judgemental, and they are great people to know. They have huge experience of student difficulties, allow you to talk confidentially about personal issues, and can suggest next steps if you wish.

4.7 Equal opportunities, harassment, gender and racial issues

All universities aim to be inclusive, treating everyone, students and staff, fairly and justly. Happily, most of the time there are few problems. This section cannot discuss these issues in detail but if you feel there is a problem, please seek advice. Many staff have equality and diversity training. Talk to someone sooner rather than later, the Students Union is a great starting point. The NUS (2018) statement 'it's about making sure that everyone feels able to participate' sums up the commitment very ably. Your problem is unlikely to be a new one. There is a great deal of advice and experience available, tap into it. Your university will have support similar to that shown on the Cardiff and Cambridge web pages listed here.

Cardiff University 2018 Equality and diversity, https://www.cardiffstudents.com/advice/general/equality/ Accessed 16 January 2019.

NUS 2018 Equality and Diversity, https://www.nusconnect.org.uk/learning-academy/equality-and-diversity Accessed 16 January 2019.

University of Cambridge 2018 Equality, diversity and inclusion, https://www.studentwellbeing.admin.cam.ac.uk/equality-diversity-and-inclusion Accessed 16 January 2019.

4.8 Stress and anxiety

Stress is a normal bodily reaction to the demands of daily life, and arises when you feel that life's physical, emotional or psychological demands are getting too much. Some people view stress as a challenge and it is vital to them for getting jobs done; stressful events are healthy challenges. Where events are seen as threatening, then distress and unhappiness may follow. If you are feeling upset, or find yourself buttering the kettle and not the toast, try to analyse why. Watch out for circumstances where you are stressed because of:

- Expectations you have for yourself. (Are they reasonable at this time in these circumstances?)
- Expectations of others, especially parents and tutors. (They have the best motives, but are these reasonable expectations at this time?)
- Physical environment – noisy flatmates, people who don't wash up, wet weather, hot weather, dark evenings. (What can be done to ameliorate these stresses?)
- Academic pressures, too many deadlines, not enough time to read. (Can you use study buddies to share study problems? Would it help to talk to someone about time management?)
- Social pressures, partying all night. (Are friends making unreasonable demands?)

How do you recognise stress in yourself or friends? Watch for signs like feeling tense, irritable, fatigued, depressed, lacking interest in your studies, having a reduced ability to concentrate, apathy and a tendency to get too stuck into stimulants like drink, drugs and nicotine. So that covers most of us; DO NOT GET PARANOID. PLEASE do not wander into the health centre waving this page, and demand attention for what is really a hangover following a work-free term. Thank you.

Managing stress effectively is mostly about balancing demands and desires, getting a mixture of academic and jolly activities, taking time off if you tend to the 'workaholic' approach (the workaholic student spends nine hours a day in the library outside lectures, has read more papers than the lecturer and is still panicking). Get some exercise – aerobics, Zumba, line dancing, jogging, swimming, any sport – walk somewhere each day. Practise relaxation skills – Tai Chi classes are great fun. Study regularly and for sensible time periods. Break down big tasks into little chunks and tick them off as you do them. If you are stressed, TALK TO SOMEONE.

See your university website or start with:

University of Leicester 2018 Stress management for presentations and interviews, https://www2.le.ac.uk/offices/ld/resources/presentations/stress-mgt Accessed 16 January 2019.

University of St Andrews 2018 Managing stress, https://www.st-andrews.ac.uk/students/advice/leaflets/stress/ Accessed 16 January 2019.

4.9 Specific learning difficulties – dyslexia, dyscalculia and hyperactivity

Many students at university have specific learning difficulties (SpLDs), with dyslexia or dyscalculia particularly common. If you had support for dyslexia and dyscalculia through school, bring the assessment report to university. If you are newly aware of difficulties, get assessed by your university. The read&write (Texthelp 2016) or similar software is a brilliant help for many students; hearing books and papers read to you is a neat break from reading.

> Modified assessment provision (MAP) gives equality of access to students who are disadvantaged in timed assessments.

Most lecturers understand that most dyslexic students work harder than their colleagues, are more organised, take more time over writing, reviewing and reading, and take time to think about what they want to say. It is normal for the students to ask other people to help with proofreading an essay, and to be good with spell checkers and grammar checkers. One definition of graduateness is taking responsibility for your learning, which is usually very evident in the way dyslexic students organise themselves.

Attention deficit (hyperactivity) disorder (ADHD) needs thoughtful attention. Reflection (Chapter 2) and planning time (Chapter 3) is helpful. If you think you might forget to take your medication, please alert someone you live with so they can remind you if necessary.

Remember that the vast majority of students with specific learning difficulties graduate and get graduate jobs.

Check for resources on your university site, or these (all websites accessed 16 January 2019).

British Dyslexia Association 2018 http://www.bdadyslexia.org.uk/.

Dyslexia Help 2018 www.dyslexiahelp.co.uk/.

Hargreaves S and Crabb J 2016 *Study Skills for Students with Dyslexia: Support for Specific Learning Differences (SpLDs)*, Sage, London.

Manchester Metropolitan University 2018 Disability Service, https://www.mmu.ac.uk/academic-services/studentservices/learner-development/about-us.php.

Texthelp 2016 read&write, https://www.texthelp.com/en-gb/products/read-write/.

Wallbank AJ 2018 *Academic Writing and Dyslexia: a visual guide to writing at university*, Routledge, Oxford.

4.10 Mental health difficulties

Mental health issues need professional help. Your doctor, the university health service, and disabilities service are good places to start. Amongst the conditions formally recognised are anxiety, eating disorders, depression, mood swings, obsessive compulsive disorder (OCD), bipolar affective disorder, personality disorders, and psychosis.

Living away from home and learning in large classes can be stressful; it is wise to let tutors know in confidence that you may need support. They can get help quickly if you have problems. Get an early appointment with the doctor so you have access to your medication.

Mind 2018 https://www.mind.org.uk/ Accessed 16 January 2019.

Students against depression 2018 http://studentsagainstepression.org/ Accessed 16 January 2019.

4.11 Gambling

Gambling is often a hidden issue, but it is estimated that in the UK over 100,000 students have gambling debts, and 10 per cent of undergraduates miss some teaching sessions because of this habit. It is very easy to start online gaming, and for some people ruinously expensive as gambling debts take all your student loan and lead to serious bank loans. Bingo, betting machines, casinos and online gaming are legal activities, it is your (grown-up) decision to play. The stress involved in trying to manage increasing debts is known to damage students' academic success and well-being. Two starting points for information are:

Royal College of Psychiatrists 2018 Problem Gambling, https://www.rcpsych.ac.uk/mental-health/problems-disorders/problem-gambling Accessed 16 January 2019.

University of Warwick 2018 Gambling, https://warwick.ac.uk/services/counselling/informationpages/gambling/ Accessed 16 January 2019.

4.12 Dropping out

'I've been here six weeks, no one has spoken to me and I hate it, I'm off'. Happily, this is a rare experience, but every year there are a few people amongst the thousands taking GEES degrees who are not happy. The main reasons are *'wrong subject choice, I should have done …'*; *'Everyone else is cleverer than me'*; *'the course was not what I expected'*;

'I was so shattered after A level and school, I really need a break and a rest'; and *'I really felt I didn't fit in and it wasn't right for me'*. Homesickness is usual. All departments lose students for these kinds of reasons; you will not be the first or last student feeling unsettled. Tutors, and people in your Students Union will advise you. Talk to someone as soon as you start to feel unhappy, waiting will probably make you feel worse.

The good news is that most people survive the first weeks, get involved with work, social and sporting activities, and really enjoy themselves. Remember, there are 6,000–50,000+ people on campus, 99.9 per cent are very nice and at least 99.7 per cent feel as shy as you do. If you are really unhappy explore the options of suspending your studies, taking a year out or transferring to another university. Take the time to consider options, don't rush your decision.

Critically question all unsupported statistics!

4.13 Helpline web links

Most people have a great time at university, but if you or your friends have problems, here are some starting points. Non-UK students should search for equivalent organisations using key words, although the information from these sites may prove helpful. All sites were live in January 2019.

Alcohol Concern	https://www.alcoholconcern.org.uk/
Drug Wise	http://www.drugwise.org.uk/
National Sleep Foundation	https://sleepfoundation.org/
NUS Health and Well-being advice	https://www.nus.org.uk/en/advice/health-and-wellbeing/
Samaritans	http://www.samaritans.org/
Mind 2019 seasonal affective disorder	https://www.mind.org.uk/information-support/types-of-mental-health-problems/seasonal-affective-disorder-sad/#.XlJ1HSj7SAs
Social Anxiety	http://www.social-anxiety.org.uk/
The Terrence Higgins Trust, HIV and Aids Charity	http://www.tht.org.uk/
Womens Aid	http://www.womensaid.org.uk/

Words in geo-words 1

Set the stopwatch for five minutes on your phone, tablet or computer. How many words of three letters or more can you find in **SEDIMENT**? Thirty is a good score. Answers in Chapter 28.

Random useful stuff

5.1 Why Twitter?

It is scary talking in a class of 300 but everyone can put their thoughts into a Tweet and see what others think. Some lecturers use Twitter in-class because it encourages participation, sharing ideas, conversations and the development of networks. These discussions may occur at other times, which is brilliant for those who cannot be on campus. Twitter is a great place for brainstorming ideas (Chapter 17).

To get started: download the app; create your own Twitter Handle, normally @ followed by your name (e.g., @jamesd or @spike21), and you are ready to send and receive messages. Messages include a # tag so other people can find them. A module may be #ENSC102 or #GEOL2052 … Retweeting a message sends it to others. You build your network by 'Following' people. Search for the latest Twitter tips. It takes less than 15 minutes to set up and get Tweeting.

A few modules use Twitter to link up with students in other universities and countries. If you're involved in volunteering, university societies and sports activities, Twitter enables you to keep in touch with group and team members, and promote events within and beyond your institution.

No doubt social media platforms for academic discussions will evolve, but the advantage of being able to work and share information quickly will continue.

5.2 Why LinkedIn?

Some universities prompt their students to set up a profile on LinkedIn early in their degree to encourage an understanding of professional work processes and longer-term employability. It is a great place to get advice on all aspects of work and business, and the international dimension is a serious advantage to those looking for opportunities abroad. Recruiters view profiles when searching for interns and when you apply for jobs. A smart, clear piece about yourself is a big help. Most people post relatively brief information because it's a professional contact and networking site, not Facebook. For models, look at other students' sites, and those of your tutors.

As with other social media platforms, LinkedIn is updated regularly. Search online for 'LinkedIn for students' to get the most recent advice on setting up your profile. Use the website to search information about jobs, internships, starting your own business and millions of other topics. Aim to have a profile by the end of term.

5.3 Feedback – some thoughts to prompt discussion

Feedback is probably the most misunderstood term in higher education. This short section should help you to understand what it is, and how to use it. It's a good topic to raise in tutorials. Get your tutor's perspective.

Feedback is where most learning happens. It's where conversations (written or verbal) allow you to understand your performance, and identify future activities that enable you to develop further. Some of these conversations will be with tutors, but many will be with fellow students and people outside education. Take the view that *all conversations provide feedback that you can act on.*

There is a general misunderstanding that feedback is provided by lecturers to justify the mark on an assessment. Good feedback is so much more than that. Feedback may be attached to an assessment, but this would be better described as feedforward, because the information is focused on explaining what to do next to get higher marks. The reason to collect, read and discuss feedback on an essay or exam is to understand what to do next time. You can't change the mark, but you can evolve how you work and how you present your thoughts and ideas in future. After an assessment has been marked, a lecturer may start the next session with feedback. By identifying the successes and problems, the messages feedforward into improving everyone's performance in the next assignment, module or project. Lecturers want to award high marks.

Staff engage you in academic conversations face-to-face, in tutorials, online blogs, by email and Twitter, but they don't explicitly label these as 'feedback'. A tutor that comments on a student's ideas for a dissertation, is giving feedback. Advice is a form of feedback. Think of feedback as 'dialogue'. It's a process where you are an active participant in the cycle of learning. Your tutor gives you information, you respond to that. Tutors recognise that people join the learning cycle at different stages, so their advice aims to support you to learn from your current position.

Another way to think about feedback is as 'coaching'. Tutors are providing guidance and expertise to help you achieve. The conversational focus will vary with circumstances, developing your skills, your knowledge of specific topics and your longer-term life and career ambitions. Search for 'mentoring' and 'coaching' for more thoughts.

In summary, feedback may be written, online or verbal, given to you as an individual or to a group, immediate or delayed, and from a wide range of people. The trick is to take some time to absorb the ideas and decide what to do about them. Reflect and act, make time in your diary (Chapters 2 and 3). Have another look at **Try This 1.3**, p 16. Be active with feedback.

5.4 Office hours

In some universities academic staff are available to see students at set times called 'office hours', as well as being available by email. Academic staff usually tutor 30 or more students and teach different years and courses. They are busy people who will not, therefore, reply to email immediately. During advertised Office Hours they don't have teaching and meetings so you can talk together in person, or possibly by phone or Skype. Times are usually advertised on office doors, in emails or in the handbook. Making an appointment is sensible, and very helpful if you want advice or feedback on your work because your tutor can prepare. Tutors provide advice on your work in general, on your progress and other matters including directing you to other experts. If a tutor does not offer advice on 'the answer' it's probably because they are giving you the time to think and work it out for yourself. It may be frustrating for you but it is good teaching.

If you have a complaint or urgent worry, contacting your tutor is the best idea. Most problems are misunderstandings that can be cleared up very quickly, or referred to the right people for action.

5.5 Software and apps

The apps mentioned here are live in 2018–19. Some will disappear quickly, replaced by newer and smarter alternatives. Search 'apps for higher education' to see what else is recommended. Set privacy access appropriately, and use **free resources**.

Class and fieldwork research support is rapidly evolving with apps and web-pages providing accessible support via tablets and smartphones, provided Wi-Fi is available. Where possible, download apps and resources to use independently in areas which do have good coverage and remember that you are relying on having enough battery life. Digital support isn't as reliable as a pencil and notebook.

Most universities have an app link to the VLE, the Virtual Learning Environment (VLEs may be called Course Management System – CMS; Learning Management System – LMS; BlackBoard; Moodle; Student portal …). The VLE gives you access to module and timetable information, lecture notes, PowerPoints, the library, assessment information and your exam results. It's where module leaders organise online discussion groups, put answers to the questions you ask, and so on. Whatever your system is called, save oodles of time by completing the introductory online tutorials. Being familiar with how your VLE, portal and apps work will make life much easier.

Find out what software is available to you, free or at a discounted price. Try the website, the library, or IT helpdesk. If you're commuting to university each day then software to enable group work and to contact your tutor is important. **Skype, Messenger, FaceTime** or …

✓ **CamScanner** allows you to take a photo of a page, digitising like a scanner. The photo is converted into a PDF or JPEG file that you can store, print and share.

✓ **Due** and **Studious** are two of many apps that will help you track deadlines, set reminders and set off alarms to prompt action (Chapter 3).

✓ **Mint** and **Splittable** are just two of many apps that will help you keep track of your money. Splittable started out as an app to split bills between people.

✓ **OneNote** you can use in class to record discussions (make sure everyone there is happy to be recorded), and write notes with the stylus or pen. Check costs and compare with Evernote costs and function.

✓ **Evernote**, like OneNote, will help you organise notes, audio, photos, pdfs and group work.

✓ **TED** talks all have brilliant material, and are great to listen to as you walk, run …

✓ **Uber** is cheaper than taxis – keep safe getting home late at night.

✓ **Oxford English Dictionary** is useful if you are travelling with no Wi-Fi.

✓ **Dropbox** get the version for your computer and the mobile app so you can share, read and edit documents, photos and videos anywhere.

Apps for fieldwork

Every country has its own **Weather** apps. Download and use daily for safety reasons. It is as important to know when the weather will be very hot and you need extra sunscreen, a hat and hydration, as when to take cold and wet weather gear. Most cities and regions have tourist information sites, maps and other details on websites and as apps. Type Toronto (or any city name) into the search function for the App Store and explore the options. Brilliant for bus, train and subway timetables and local maps. It might be useful to have:

✓ **Compass** apps – lots of choice here.

✓ **Calculator** apps – essential for speedy calculations.

✓ **Clinometers** – for measuring slope angles. Various options in the App Store. This feature may be available on your compass app.

Apps for fun

☺ The **Logo Quiz** app tests your broad knowledge of countries and is fun to use with teams.

☺ **Geocaching**, which involves treasure hunting, orienteering and a good walk is a great way to explore a new area. It's a fun, team activity too.

☺ **Grind GIS What is geography** (http://grindgis.com/geography/types-of-geography_know-everything), described by a student as *'having all the detail needed to explain my degree to my grandmother'*.

☺ **Kahoot** for quiz-style learning.

☺ The USGS Water Science School has some true/false quizzes, and 'How much water does it take to grow a hamburger?' at https://water.usgs.gov/edu/activity-tf-properties.php.

5.6 Some data thoughts

Accuracy and precision

Data is usually noisy, imprecise, inconsistent and possibly biased. The trick is to recognise the sources of errors, acknowledge them and discuss their impact, for example in the methodology or discussion sections of a report.

Discussing error is not a matter of mega *mea culpa*, 'it was his/her fault' statements. Unless exceptional care is taken, items, questions and variables are forgotten; data collection methods influence the measurement values; and instruments may be inaccurate. You aim to minimise errors, but only the infallible succeed. Be as objective as possible. A true value, the absolutely accurate, correct value of something can be hard to determine; more often one has the best estimated value which can be described in terms of its accuracy and precision.

> **Accurate data:** measurements that are as close to the right value as possible.
>
> **Precise data:** can be reproduced consistently but are not necessarily accurate.

A grid reference position on the ground can be located with increasing accuracy as the scale of the map increases. On a 1:50,000 map, a position may be in error by ±32m, whereas on a 1:25,000 map, the error may be ±15m. The accuracy of points located from satellite imagery will depend on the height and type of the platform and the spatial resolution of the images, and biased by topographic distortions and off-nadir angles during scanning. The accuracy of GPS (global positioning system) data depends on the sophistication of the instrument and the number of satellites in view.

Reporting error

It is good practice to quote a value with its associated resolution, for example with error bars and scatter plots, giving the reader a feel for the precision and accuracy of the information. A location may be given as X ± 16m. An average, 72.25 per cent, may be reported with the range 66.3–78.2 per cent. Temperature data from a thermometer, which can, at best, be read to 0.2 of a degree, would have the uncertainty associated with a reading expressed as $8.2 \pm 0.2°C$. Measuring temperature with a thermocouple should give greater accuracy, because the measurement can be made to the nearest $0.01°C$. It is a more precise measurement: $8.23 \pm 0.01°C$.

However, if the thermocouple consistently under-reads the thermometer, although the data are more precise, they would be less accurate than those from the thermometer. The readings would be precise but inaccurate, due to instrumental error. Similarly, if a GPS is not properly calibrated, subsequent readouts are consistent but inaccurate.

Decimal places

A disadvantage of digital technology is that calculations can be reported to many decimal places. Use what is relevant for your study or instrument. Avoid the absurdity of 'a family with 2.45678 children', or, 'Therefore we conclude the residents of Hobbiton make 3.21776 shopping trips per week to Buckland', or 'Survey results show the population of Stats-on-Sea are 26.3732 per cent Hindu and 56.4567 per cent Christian'.

How many decimal places to quote? Think about the accuracy you require and the application. An average temperature of 16.734567°C is experimentally accurate and appropriate for a physicist doing neutron experiments; for a soil temperature, 16°C or 16.7°C will do. A digital pH meter used in the laboratory or field may show two decimal places, but the digits are unlikely to settle because pH is not a stable measure. In practice, a soil pH will have local variability and to report to more than 0.2pH has no practical scientific value.

Bias: tendency to express one view or opinion without presenting alternative views.

Statistical bias: occurs in a number of ways – especially when data is selected such that not all samples could be equally selected, or a hypothesis is tested which is not independent.

Is the sample biased?

Bias is a consistent error in data. There is plenty of information in statistics, laboratory and fieldwork sessions about taking samples in the right way and with enough replicates so that bias is minimised. Think about potential sources of bias when considering results. I think you would agree that sampling two midday temperatures in July at Filey in 2020, and using these data to describe the average climate of the UK in the twenty-first century would leave the reader unimpressed. Such data are obviously inadequate and biased. Does your dataset have similar, albeit less blatant, drawbacks?

Data entry errors

These are very, very, very, very common. Check your data very carefully every time against original field or laboratory notes. Using the spreadsheet command to show the smallest and largest datum can help to find the commoner errors: forgetting a decimal point or typing two items together as a single entry.

Computational errors

Have all the formulae for calculations been entered correctly? Cross-check the equations and look at the answers to make sure they are in the right 'ball park'. A quick, back-of-the-envelope calculation using whole numbers can give a feel for the range of answers to expect. If answers fall outside the expected range, check all the calculation steps carefully. Keep units the same as far as possible, and make careful notes when the units change through different stages of a calculation.

5.7 Equations, chemistry and maths!

The education you were given from ages 5 to 16 comes into play at university. If some things, like mathematics and chemistry, are a bit dormant, here are some basics. **DO NOT PANIC** (well, only a little). Scientific notation is shorthand. You can do the easy ones: EITHER a lecturer says: 'Water added to a catchment from precipitation, could be accounted for by evaporation loss, evapo-transpiration loss, runoff in the river, and changes in the soil and groundwater stores' (28 words); OR

$$P = E+ET+Q+\Delta S+\Delta G \text{ (not 28 words)}$$

It is essential to define the elements of the equation, but having done it once, time and space is saved later by not having to write it all out again. Where appropriate, WHICH IS 99.9999 per cent OF THE TIME, the units of measurement are included. Darcy's Law for water flow through a soil or sediment sample:

Q = KIA where	Q = Discharge of water though the soil K = Hydraulic conductivity of the soil (a function of water content) I = Hydraulic gradient A = cross-sectional area of the soil sample

Gets about 60 per cent in an exam. Top marks, instead, for:

Q = KIA where	Q = Discharge of water though the soil K = Hydraulic conductivity of the soil (a function of water content) I = Hydraulic gradient A = cross-sectional area of the soil sample	$cm^3.s^{-1}$ $cm.s^{-1}$ dimensionless cm^2

Chemistry

The following elements and compounds, frequently starring in geophysics, mineralogy, geomorphology, hydrology and soils courses, are simply convenient, internationally recognised abbreviations:

Symbol	Name	Symbol	Name	Symbol	Name
Al	Aluminium	Fe	Iron	P	Phosphorous
C	Carbon	H	Hydrogen	Pb	Lead
Ca	Calcium	Hg	Mercury	Rn	Radon
Cd	Cadmium	K	Potassium	S	Sulphur
Cl	Chlorine	Mg	Magnesium	Se	Selenium
Cr	Chromium	Mn	Manganese	Si	Silicon
Cs	Caesium	N	Nitrogen	U	Uranium
Cu	Copper	Na	Sodium	Zn	Zinc
F	Fluorine	O	Oxygen		

$CaCO_3$	Calcium carbonate, limestone			CaSO4	Calcium sulphate
$C_6H_{12}O_6$	One of the monosaccharide sugars, glucose or fructose				
CH_4	Methane			CO	Carbon monoxide
CO_2	Carbon dioxide			HCl	Hydrochloric acid
H_2O	Water			H_2S	Hydrogen sulphide
H_2SO_4	Sulphuric acid				
NO_x	Nitrogen oxides, includes NO and NO2				
N_2O	Nitrous oxide			NO	Nitric oxide
NO_2	Nitrogen dioxide			NO_{-2}	Nitrite
$NO-_3$	Nitrate			NH_3	Ammonia
NH_4	Ammonium			O_3	Ozone
SO_2	Sulphur dioxide			SO_3	Sulphur trioxide

The knack is to read chemical formulae as sentences: 'Erosion of limestone used as building stone is caused by acidified rainfall, in the form of sulphuric acid, reacting with the limestone to form the weaker, more erodible calcium sulphate, carbon dioxide and water.' Or substitute chemical symbols for words:

$$H_2SO_4 \quad + \quad CaCO_3 \quad \rightarrow \quad CaSO_4 \quad + \quad CO_2 \quad + \quad H_2O$$

(Sulphuric acid + limestone **reacts** to give Calcium sulphate + Carbon dioxide + water)

Easy! Chemical symbols are like acronyms, a substitute for words. Remember that in chemical equations subscripts represent the number of atoms of each element of the molecule. The coefficients (large number: 12H in $12H_2O$) represent the number of molecules of the substance in the reaction. Balance equations by changing the coefficients. Never change subscripts.

Now it's your turn. Practise with **Try This 5.1**.

TRY THIS 5.1 – Chemical formulae

Read the following formulae as sentences. What do these relationships describe and what are their GEES meanings? Answers at end of chapter.

$$CH_4 + O_2 \rightarrow CO_2 + 2H_2O$$
$$NH_3 + H_2O \rightarrow NH_4 + OH$$
$$SO_3 + H_2O \rightarrow H_2SO_4$$
$$12H_2O + 6CO_2 + 709 \text{ kcal} \rightarrow C_6H_{12}O_6 + 6O_2 + 6H_2O$$

Matrix data

Life gets to be more fun when data comes in the form of tables and matrices. A table that lists all the students in your class from Sheila to … n, and the local pubs, The Lost Orienteer to … n, can be referred to using matrix algebra. Yes, algebra. **DO NOT PANIC.**

What seems to confuse some people is making a distinction between the code used to refer to positions in a matrix, and the data values themselves. Basically, it works like map reading with grid references. If the data in the matrix refers to cash spent by individuals in the pub over a term, as here:

Student		Hostelry				
		The Lost Orienteer	The Lorenz and Lowry	The Mattock and Spade	The Hazard and Impact	… n
		i_1	i_2	i_3	i_4	… i_n
Sheila	j_1	£25.50	£16.75	£0.00	£3.00	
Wayne	j_2	£45.70	£0.00	£16.80	£6.50	
Spike	j_3	£52.80	£12.80	£36.45	£22.40	
Liz	j_4	£28.67	£24.87	£16.85	£35.78	
Jenny	j_5	£45.23	£12.60	£21.76	£0.00	
…						
n	j_n					

Each of the pubs is coded as $i_i \ldots _n$ and each student as $j_1 \ldots _n$. The total number of pubs is $\Sigma i = 1 - n$ and the total number in the class is $\Sigma j = 1 - n$. You can refer to a single cell: the shaded cell for Liz in The Lorenz and Lowry is i2j4. The total amount spent by Sheila is the sum of all the columns $i = 1-n$ for row j_1. That's $25.50 + 16.75 + 3.00 = £45.25$.

Read equations carefully and remember to solve the elements in brackets first. Just as a reminder, which of these instructions is right?

$a + 15b - e$	a) Multiply b by 15, then add a and subtract e. b) The sum of 15 times b and a minus e. c) a plus 15, times b and subtract e.
$6 (x^3 + y^2)$	d) Take the square of y and add it to the cube of x and multiply the total by 6. e) Multiply the cube value of x by 6, and then add the squared value of y. f) Cube x, then square y, add the two values together and multiply by 6.
$2x^2 - y^2$	g) Two times the square of x, minus y squared. h) Square x and y, subtract and multiply by 2. i) Square x and multiply by 2, and then subtract the squared value of y.

(The answers are a, b, d, f, g and i.)

Percentage problems?

50% = 1/2 = 0.5	33.33% = 1/3 = 0.333	25% = 1/4 = 0.25
20% = 1/5 = 0.2	5% = 1/20 = 0.05	1% = 1/100 = 0.01

5.8 Dictionaries and atlases, do I need one?

Atlases are useful but expensive; the library has plenty and the internet has many alternatives. While an atlas is not a vital purchase, for some topics and field visits it's very, very helpful to be familiar with the names and locations of cities, rivers and … It is vital to check the date of the map's publication because place names change. If your Gran wants to give you something useful for Christmas an atlas is a good wheeze – it helps with planning holidays.

Your word-processing software (and the internet) has **dictionary** and **thesaurus** functions. The subject dictionaries below contain short explanations which are much more informative than a general dictionary. They can be a good starting point for research.

Allaby M 2013 *A Dictionary of Earth Sciences*, (4th Edn.), Oxford University Press, Oxford, http://www.oxfordreference.com/view/10.1093/acref/9780199653065.001.0001/acref-9780199653065 Accessed 15 January 2019.

Castree N, Kitchin R and Rogers A 2013 *A Dictionary of Human Geography*, Oxford University Press, Oxford, http://www.oxfordreference.com/view/10.1093/acref/9780199599868.001.0001/acref-9780199599868 Accessed 15 January 2019.

Mayhew S 2009 *A Dictionary of Geography*, (5th Edn.), Oxford University Press, Oxford, http://www.oxfordreference.com/view/10.1093/acref/9780199680856.001.0001/acref-9780199680856 Accessed 15 January 2019.

Park C and Allaby M 2017 *A Dictionary of Environment and Conservation*, (3rd Edn.), Oxford University Press, Oxford, http://www.oxfordreference.com/view/10.1093/acref/9780191826320.001.0001/acref-9780191826320 Accessed 15 January 2019.

5.9 Writing

It's not what you read or learn but how you articulate information in writing that is marked.

If you are having difficulty getting words onto the page or want to be a more confident writer, try these ideas. Start by considering what you do now: where your style fits with the approaches in Figure 5.1. Writing isn't easy. You get mixed up, forget what you want to say, run out of things to say, repeat stuff, forget the right words, change your ideas as you write and decide to start again. There are so many distractions. This is all normal.

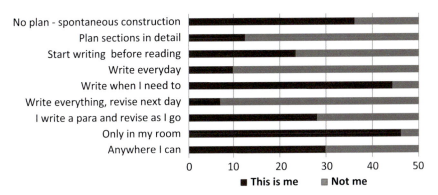

Figure 5.1 How and where do you write?

Vital tactic Recognise that surfing the web, answering emails, making coffee and phoning are all distractions that get in the way of writing. Decide that you will write for 60 (15, 45, 90) minutes on a regular basis. Turn off email, web windows, music, TV … and concentrate on writing before turning anything on again, this includes web searching, looking up references and getting another coffee.

Prepare to write Being ready to write is the key to success. Research, read, make notes, discuss ideas, sort notes into an order that makes a story **and plan uninterrupted time for writing.** Get the time in your diary, and make sure you won't be disturbed (*turn off phone, email …*).

Uninterrupted writing First, talk to yourself or a friend for about **five** minutes explaining what it is you are going to write about. This will get your ideas together and focus your brain on writing. It helps concentration. Then write for the next 30–45 minutes without stopping.

Type or talk? Most word-processing software lets you talk to your computer. You can probably talk faster than you can type, letting you get ideas onto the page relatively quickly. Ideally talk through the 'story' and then go back to edit the errors. If you correct as you go you are likely to lose track of the story. Editing carefully is very important as some words will be heard/herd/her/here/hear incorrectly.

Warm up – practise ideas These ideas will help you get started by writing short (*useful*) pieces:

- Reading and writing: Open any paper/book/online resource and read for ten minutes, (*turn off phone, email …*) and write for five minutes.
- Choose a photograph or diagram from a text/paper/field visit. Take five minutes to talk **out loud** about what it means, its value, relevance, place in what you are learning. If there is someone handy, talk to them. Then write a paragraph that explains the value or relevance of your resource.

Writing every day This sounds like too much trouble, but it will develop your communication skills and the chance to remember material from a class by reinforcing the information, which will help you to be confident in developing your own ideas and ability to be critical. Aiming for five minutes thinking aloud and ten minutes of focused writing each day (*phone and email off*) takes very little time and is a learning win.

Writing support Universities have a variety of support opportunities for writing. Look out for opportunities advertised in the library, learning support and student services areas. Some universities run writing workshops, and have writing cafés where you may find staff who can advise you.

> *The art of writing involves iteratively developing and organising ideas, making links and evolving thinking. Good writing involves drafting, refining and refining.*

Search for 'art of writing quotes' for more insights into writing. Another resource:

Gemmell K 2015 *The Write Stuff, seven steps to writing excellence*, (4th Edn.), Emphasis, Brighton.

5.10 Word processing – do you know …?

OK – you have been typing for at least ten years, but are you using all the bells and buttons to present the most professional document every time? Take 30 minutes

NOW to explore what your word-processing package can do – it will help you work smarter. Tick ✓ off the following elements most often missing in student work, and celebrate a full list (Figure 5.2).

Function	Value	
Insert page numbers	Vital in every document, please add every time.	✓
Spell and grammar checker	Set it to always be on and at the highest level. The suggestions are not always what you want but will make you think about your writing style.	
Language	Make sure your documents are spell checked in the right language. English (UK) is not the same as English (US).	
Bulleted or numbered lists	Helps the reader to understand the message. Insert jolly ♟ ✓ ✗ ☺ symbols as in this textbook too, where appropriate.	
Images	Photos, diagrams, charts, clip art; cite the source correctly, pictures usually save hundreds of explanatory words.	
Header and footer information	Important in longer documents. Put your name, student number, module code and title, in the header or footer, with the page numbers.	
Tables	This book has many tables to make information clearer. Tables are not just for numbers. Explore the Insert Table function.	
Titles and subheadings	Use 'Styles' to structure sections with larger font sizes. Subheadings help you see if your answer is balanced: a three-part question leads to three subheadings of roughly equal weight. A few tutors dislike subheadings; if so, write the essay with subheads, check for balance, then delete the subheadings, and submit.	
References	Using Endnote or similar software will help to create the reference list, keeps track of entries, and saves time when finishing documents.	
Footnotes	More often used by historical and cultural geographers than in the sciences, footnotes add more detailed explanations, but are not needed in tutorial essays.	
Thesaurus	Stuck for the right word, see what alternatives are suggested.	
Review and track changes	These commands are essential for group writing, so people can see who has added and changed each section. The track change function highlights who writes which section	
Columns	Useful for newsletter, handbill and PR writing. A two-column table will produce a similar result.	
Sort	Essential for getting references into alphabetical order in every essay. And good for all other ordering tasks.	
Read Aloud	Reads the document. Very useful when proofreading. It can be easier to hear mistakes than spot them while reading. https://www.texthelp.com/en-gb/products/read-write/.	
Line spacing, fonts, justification; colours; borders; margins … all the elements that make smarter documents.		

Figure 5.2 Checklist of word-processing skills

Always use your spell checker but 'spill Chequers due knot awl weighs git it write'.

5.11 An acronym starter list

Add acronyms to make this your own resource.

AI	Artificial Intelligence (unless it is an agriculture lecture!)
AI	Appreciative Inquiry
AONB	Area of Outstanding Natural Beauty
BOD	Biological Oxygen Demand
BPEO	Best Practicable Environmental Option
CBR	Crude Birth Rate
CDR	Crude Death Rate
CFCs	Chlorofluorocarbons
COD	Chemical Oxygen Demand
CPD	Continuing professional development
CV	Curriculum Vitae (résumé)
DEM	Digital Elevation Model
DO	Dissolved Oxygen
DSS	Decision Support System
DTM	Digital Terrain Model
ECLA	Economic Commission for Latin America
EIA	Environmental Impact Assessment
GATT	General Agreement on Tariffs and Trade
GDP	Gross Domestic Product
GNP	Gross National Product
HCFCs	Hydrochlorofluorocarbons
IMF	International Monetary Fund
LCA	Life Cycle Analysis
MCQs	Multiple Choice Questions
NATO	North Atlantic Treaty Organization
NGOs	Non-Governmental Organisations
NIABY	Not In Anybody's Back Yard
NIR	Natural Increase Rate
NSA	Nitrate Sensitive Area
OAS	Organization of American States
OECD	Organisation for Economic Co-operation and Development
OPEC	Organization of Petroleum Exporting Countries
PCBs	Polychlorinated Biphenyls
PDP	Personal Development Planning
SMEs	Small and medium sized enterprises (below 50, or 50–250 staff)
SSSI	Site of Special Scientific Interest
TCV	The Conservation Volunteers
TDS	Total Dissolved Solids

TFR   Total Fertility Rate
TLV   Threshold Limiting Value
TNC   Transnational Corporation
VLE   Virtual learning environment

5.12 What is a university for?

This is a big discussion topic with many dimensions to the answer – useful in a tutorial. It is interesting to think (reflect) about what your university experience is giving you and how your expectations and values change as you progress. Some starter thoughts:

Where a love of learning is developed	A space for self-discovery
Where assumptions are challenged	Where you can prepare for an uncertain world
Where different ideas are tolerated	Developing socially responsible thinking
Finding stuff out	Acquiring advanced knowledge
Meeting people that will be useful in life	Sharing dreams and experiences
Interacting with and learning from people	Where intellectual curiosity is encouraged
Preparing people for jobs	A fun place to be

Missing vowels 1

All of the vowels are missing from these ten recycling-related terms. Can you reconstruct them? Answers in Chapter 28.

BDGRDBL	CNVRTR	DCMPS	LNDFLL	MTHN
PCKGNG	RCVRY	RS	SPRTN	WST DSPSL

Answers: Try This 5.1 – Chemical formulae

$CH_4 + O_2 \rightarrow CO_2 + 2H_2O$ (When methane and oxygen are burned together the reaction gives carbon dioxide and water, a greenhouse gas issue.)

$NH_3 + H_2O \rightarrow NH_4 + OH$ (Describes the ammonia to ammonium oxidation – reduction interaction, a soils process.)

$SO_3 + H_2O \rightarrow H_2SO_4$ (An atmosphere interaction where sulphur trioxide and water combine to form sulphuric acid, which causes acid rain.)

$12H_2O + 6CO_2 + 709\ kcal \rightarrow C_6H_{12}O_6 + 6O_2 + 6H_2O$ (Describes the synthesis of water and carbon dioxide with light energy in plant leaves to produce the carbohydrate $C_6H_{12}O_6$ for plant growth with water and oxygen as waste products – a biogeography issue.)

6

Careers and employability

I think I want a career, but maybe what I want is a regular cheque?

The thought of leaving university, applying for jobs and starting a career is probably as far from your mind as the state of the Bolivian economy, but internships, work placement and summer job applications come round quickly. A focused CV that demonstrates your professional skills and commitment to development can significantly increase the chance of selection.

OK, OK, I agree that getting a job is not a Study Skill, but having great skills will get you a professional job. What you need is the skill to explain your skills to the employer who wants your skills. Knowing what you want to do after you graduate is difficult and there is an enormous choice available. Your first advice point is your careers centre, or a web search for *careers + service + university*. Make time to think about what you enjoy, find rewarding, what you dislike and ask, 'How does what I know about myself, my skills and my attitudes, affect what I want to do when I graduate?'

> *I went to uni, had a lovely time, and am back home with mum and dad. It's OK but makes your degree seem a bit worthless …*

This chapter mostly references UK sites for researching career options – I'm sorry there is no space to reference all the parallel international organisations. They exist, check them out. The Country Profiles, on the Prospects (2019) website, are great starting points for information about employment in countries from Australia to the USA.

The *professional* skills employers seek that you will enhance during your degree include:

✓ communication, both written and oral, and the ability to listen to others;
✓ interpersonal (social) skills, the capacity to establish good, professional working relationships with clients and colleagues;
✓ organisation, planning ahead, meeting deadlines, managing yourself and co-ordinating with others;
✓ problem analysis and solution: the ability to identify key issues, reconcile conflicts, devise workable solutions; clear logical thinking; prioritising and working under pressure;

✓ clear thinking, intellectual skills, judged by how effectively you translate your ideas into action;

✓ leadership skills, many graduates eventually reach senior positions, managing and leading people;

✓ teamwork, working effectively in formal and informal teams;

✓ adaptability, being able to initiate and respond to changing circumstances, and to continue to develop one's knowledge, interests and attitudes to adapt to changing demands;

✓ technical capability, the capacity to acquire appropriate technical skills, including scheduling, information technology (IT), statistics, computing, data analysis, and to update these as appropriate;

✓ setting and achieving goals for oneself and for others.

OK, this is a long list. Well-planned degrees have a good mix of all of these activities across the whole programme. By the time you graduate you should feel confident in listing all these skills and more on your curriculum vitae (CV), and be able to explain where in your degree these abilities were practised.

6.1 Jobs, employment and/or a career: start now

Ignoring employment advice and research at university risks you returning home to continue with vacation work (at vacation pay) rather than kick-starting the rest of your life with new opportunities. Researching careers will reveal exciting summer vacation and gap-year work opportunities. You know that all GEES graduates have brilliant work-related skills, BUT most employers are ignorant of the skills your degree has given you, and don't know you either. Remember that your degree makes you very employable. Most graduates are making the most of their skills in careers that are not degree-related. Pursuing certain careers requires further training (management consultancy, IT developer, solicitor, nursing, catering), but it is all good fun, interesting and usually you are paid while you train. Look at the literature in your campus careers shop, on the web, and don't rule anything out at the start. Having selected some options, research them properly. Fully prepared, you will stun employers with a well-written CV (curriculum vitae) and in interviews.

Careers advisers will suggest that you establish a professional profile on LinkedIn as part of preparing for graduation. It is a good space for networking and employability advice. Setting up a profile here is straightforward and will **help you** focus on options for the future (see Chapter 5.2).

I didn't go to the careers centre because that would have meant making decisions, but being home doing my summer job full time is a bit rubbish really.

✓ Start career thinking NOW – it will help you focus on vacation opportunities.

✓ Set aside some 'Career Thinking Time' each fortnight. Half an hour or so on the bus, in the bath …

✓ Ask people, family, friends about what they do, and consider whether a similar role would suit you.

✓ Drop into your careers centre every couple of weeks and see what is happening. Go to their employer events. Keep track of career possibilities in the back of your work file. Make notes as you go.

✓ Create and update your CV.

✓ Consider Figure 6.1. What are your thoughts and ambitions? What do you want?

Figure 6.1 Students' career thoughts, what are yours? (Wordle created using WordArt.com.)

Sorting out the opportunities

Very few people stay with one organisation for life; changing every three to five years is normal. Start making notes of your 5-year, 10-year and 15-year plans. Depending on your ambitions, research what you need to do now to make something possible in the future. For example, if you want to work in the charity sector on water supply problems, then plan to take an engineering hydrology MSc and work with a firm of consulting engineers before moving to work with Water Aid. If you plan to start your own business, then working in one or more large companies to get management skills and experience of different aspects of business, and completing a part-time MBA might be good. Potential professional geophysicists will need a Master's degree to get their first job. This type of planning helps you acquire practical professional experience and a network of contacts to support you in the future. Use **Try This 6.1** to focus your thinking.

TRY THIS 6.1 – My ambitions

What are your professional AIMS and ambitions?
- **?** Do you want to be in a big organisation with structured training, options to change direction and plenty of colleagues?
- **?** Do you want the security of a well-known organisation?
- **?** Do you want to work for yourself?
- **?** Where do you want to live?
- **?** Does being part of a small or medium-size enterprise (SME) appeal?
- **?** Have you already seen your ideal job?
- **?** What sort of lifestyle would you like (pay, holidays, location …)?

Try This 6.1 poses difficult questions; create the time to think about what will suit you. Your search for a career is just the first step. Some people opt for 'portfolio careers', mixing a series of remunerated and unremunerated appointments and fitting them around other opportunities, including family and leisure activities (Hopson 2018; University of Kent Careers Advisory Service 2019). Tackle **Try This 6.2**.

TRY THIS 6.2 – Assessing your skills and experiences

Ask your friends and family about the careers they think you will enjoy, and to describe the skills that they know you have. Think about activities and experiences that motivate you. What work experience should you seek? What makes you stand out?

Make some notes about your ambitions and your skills. Keep these notes safe to read before interviews.

6.2 Research your career options

Knowing what you want to do when you graduate is difficult and there is an enormous choice of careers. Tieger and Barron-Tieger's (2014) *Do What You Are* can be a good start. Environmental scientists and everyone interested in environmentally minded opportunities should start with YouTube and the Work2savetheworld (2011) video clips. Look at Carolyn Roberts explaining how to be a water scientist and work with the police to track dead bodies in rivers; at Huw Griffiths, a marine biologist working in the Antarctic; at Katharine Robinson, a sustainability consultant – just three of a selection of very useful insights from people doing their jobs.

Geologists start at the British Geological Society of London (2019) 'Careers in the Earth Sciences' website and chase through the links. There are links to studying earth sciences and to voluntary work too. Geographers start with the Royal Geographical Society (2019) 'Careers with Geography' website. Chase up YouTube and the Work2savetheworld (2011) video clips.

Fancy teaching in infant, junior or high school? In this career you can set up and enjoy running field trips to the locations of your choice. The UCAS (2019) Teacher Training website details UK courses.

For further study there are MA or MSc opportunities around the world. Look at website options in planning, civil engineering, information science, computing, management, sociology, politics, international studies; or applied courses such as conservation, water management, applied meteorology and climatology, urbanism, tourism, regional science, environmental planning, society and space … If you fancy three years' in-depth study of one topic, then research PhD opportunities. Ask the postgraduate officer in your department for information and advice.

Get organised. Use **Try This 6.3** and **Try This 6.4** to help you to evaluate your options.

Mentoring and coaching schemes

Some universities have schemes where alumni (graduates of your university) and others offer a mentoring or coaching programme which provides insights into professional workplaces, a chance to develop your network of contacts and practise networking. Mentoring has many different aspects and is used in different ways. Talking through your career aspirations with someone in the workplace is helpful. A good mentor will try to understand exactly what you hope to achieve and help

---

⚙ TRY THIS 6.3 – Researching careers

Being systematic in researching opportunities helps. Whether in your careers centre or surfing the internet:
Pick a career or occupation you think you might enjoy, perhaps using the planner in Prospects (2019). Then:

1. Make a list of the things you imagine the job will involve.
2. Find job descriptions and literature and/or websites for three companies/ organisations – what skills do they need? (Start at the occupational profiles on Prospects (2019) which contains graduate case studies.)
3. Make a list of the skills and the evidence you can quote to show you have these skills.
4. How did your original ideas about this occupation match your first ideas? Are there surprises?
5. Is this the right type of company for you? How do your skills and aspirations fit the package? Do you want to do the same job but in a larger/smaller company?

Now research another career. Back to point 1.

> ⚙️ TRY THIS **6.4 – Unlikely options**
>
> It really helps to confirm your job or career choice if you know what to avoid.
>   Pick two employment opportunities that you feel are not right for you and check out the websites, start with Prospects (2019). Many people are surprised when they see what different organisations and careers involve and realise that these are potential options. Equally, you may not like what you find and confirm that this is not a career for you.

you to get there, but will also challenge you to try new things and move out of your comfort zone. Check out your careers service office, and the many websites that offer support under the headings of mentoring and coaching.

6.3 Making applications

The CV

Although most applications will be made using online forms, you need a succinct, accurate, professional CV (curriculum vitae) and a covering letter handy so you can copy and tailor material onto the forms. The CV element informs employers about your past and current qualifications and experience. The covering letter links to your CV; use it to say more about the skills and experience that make you the ideal person for the job. ALWAYS revise your CV and letter for every application to match all the requirements (Figure 6.2). Keep a source CV file with all of your information to make it easy to put together a targeted online application.

The employer reading your CV does not know about you or your skills. Put up front your personal, transferable, professional and academic skills. While at university you have: given 20+ presentations to groups of 5–50; used PowerPoint; become comfortable and skilled with spreadsheet and word-processing packages, GIS and mapping packages, modelling, statistics, and hands-on laboratory skills; and taught yourself to use unfamiliar software. Updating your CV throughout your degree (after every semester) saves time later.

Tailoring your CV to each job is vital (see Figure 6.2, **Try This 6.5** and **6.6**). Most advertisements list the essential and desirable skills expected; make sure that your CV and letter addresses each one using these skills as subheadings. For post-graduate and research posts, focus on your geographical, earth or environmental sciences skills; for generic jobs focus on transferable skills and the work and student experiences that are particularly relevant. Most importantly for a written CV:

✓ two sides of A4 only;

✓ professionally presented – show off your word-processor skills;

✓ punchy and precise – no waffle please;

✓ consistent in explaining what you have done – don't leave time gaps;

✓ consistently ordered within sections – put the most recent activities and experiences first.

Be very careful to treble-check and spell check every entry before sending online applications.

Covering letter

Covering letters should be short, positive and relevant. State which job you are applying for, why you are interested and the skills and experience that you can offer. Show that you are a professional, career-minded person by stating how you see your career developing. Expand on a couple of points from your CV that are highly relevant for the job: 'As you see from the CV, I have considerable experience of …' Don't waffle or repeat chunks of the CV. Get to the point and keep it short.

Personal details	Full name, address, email, telephone number (UK Equal Opportunities legislation means date of birth, nationality and marital status are optional).
Education	University name, programme details, degree result and, if relevant, brief mention of projects/modules undertaken. High school name and results from final exams. For earlier years summarise briefly.
Work experience	Work placements, summer jobs, voluntary work, any permanent or part-time work. Place these in reverse order, the most recent first. For the most recent and relevant jobs, use two or three sentences to state the content or skills you used and developed. Be concise but relevant. For example, if you're applying for a planning job, then your CV might contain a section headed 'Planning Experience' which showcases paid and voluntary planning experience that you acquired during your degree, your work or your sports and hobbies.
IT skills	Highlight this area. Remember the packages used in practicals and projects, statistics, mapping, GIS, programming, Excel, Access, Word, PowerPoint, internet searching, electronic library skills and … You may not feel very expert, but many applicants have fewer skills. Explaining you have taught yourself to use software (such as PhotoShop) is valuable information for an employer – their interest is that you can teach yourself to use software.
Other skills	Include driving licence, first aid, and language skills. Explain your transferable skills using examples rather than lists: '… showed that I can be persuasive and present my case well'. 'Being left to manage the … on my own was a challenge that was daunting at first. But I was able to cope with all the demands that were made and reorganised my work to meet the changing needs of the clients'. 'I enjoyed solving problems, negotiating and aiming to ensure the outcome was satisfactory for everyone'.
Interests and activities	Be honest and don't just make a list. 'I play tennis for the university, rugby for the second XV hall team and have enjoyed coaching juniors at my local school'. 'I play bridge for a local club, where I co-ordinate transport to matches'. 'At present I prefer to concentrate on aerobic training for skiing but intend to play more squash in future'.
Referees	Cite two, one academic and one from work. Always ask permission. Giving referees a copy of your application helps them to write a focused letter on your behalf.

Figure 6.2 Contents of a CV

⇨ TOP TIPS

✓ Revise your CV and letter for every application to completely match the requirements.

✓ Always write in formal English. This may seem obvious but most applications are submitted electronically where it is easy to drop into a casual, email writing style. (V. bad idea.)

✓ Draft your CV and letter, then pretend to be the human resource manager reading it. Does it imply an enthusiastic, energetic and creative person with oodles of skills? Redraft until it does.

Think of yourself as a marketable product. You possess many skills that employers are seeking; it is a matter of articulating them clearly to maximise your assets.

The interview

The challenge is to prepare properly, otherwise your effort with application forms and travelling is wasted. Look at video clips of interviews for advice on what to do and not to do. YouTube has some great examples. Read and reread company literature; read everything they send; read and reread the advertisement; map your skills to their requirements. Have more examples ready. Most questions are straightforward; the trick is to have thought about the answers. Start preparing answers to:

☺ Why are you choosing to be a …?

☺ What can you offer to …? (human resources, marketing, software sales)

☺ What prompted you to apply to this organisation/sector/company/country?

☺ What do you know about this business/sector?

☺ What interests you most about this job?

☺ What particular skills do you have which will help you make a success of this job?

☺ Why do you want to work in a large/middling/small organisation?

☺ Why should we offer you a job with us?

☺ Where do you see yourself in three years'/five years' time?

☺ What were the three most significant/tough/traumatic things that happened in your year abroad/life? How did you manage them?

☺ Variations on: What are your strengths and weaknesses? What can you offer the company?

An interviewer for most commercial jobs has no interest in your knowledge of coasts, boreholes or rural development. But he might want to know that

⚙ TRY THIS **6.5 – Action words for applications and CVs**

Draft your CV and letter, then consider whether you are really expressing your skills fully. Try CV bingo – how many terms can you SENSIBLY include?

Achieve	Catalogue	Delegate	Forecast	Market	Produce	Retrieve
Accomplish	Chair	Demonstrate	Guide	Mediate	Promote	Review
Act	Clarify	Design	Generate	Meet deadlines	Propose	Revise
Adapt	Classify	Determine	Handle	Moderate	Provide	Schedule
Allocate	Collaborate	Develop	Help	Modify	Publicise	Screen
Advertise	Collate	Devise	Identify	Monitor	Publish	Select
Analyse	Collect	Direct	Implement	Motivate	Purchase	Serve
Advise	Communicate	Distribute	Improve	Negotiate	Qualify	Set up
Apply	Compare	Draft	Increase	Obtain	Raise	Solve
Administer	Compile	Edit	Influence	Operate	Recommend	Sort
Appraise	Complete	Employ	Initiate	Order	Record	Staff
Approach	Compute	Encourage	Innovate	Organise	Recruit	Summarise
Approve	Conduct	Ensure	Install	Oversee	Redesign	Supervise
Arrange	Configure	Establish	Instruct	Participate	Reduce	Support
Assemble	Consult	Estimate	Integrate	Perceive	Regulate	Survey
Assess	Contract	Evaluate	Interpret	Perform	Renew	Synthesise
Assign	Control	Examine	Interview	Persuade	Reorganise	Teach
Assist	Co-operate	Execute	Introduce	Plan	Repair	Train
Attain	Co-ordinate	Expedite	Invent	Prepare	Report	Transmit
Audit	Correct	Facilitate	Investigate	Present	Represent	Update
Budget	Create	Familiarise	Lead	Prioritise	Research	Validate
Calculate	Decide	Formulate	Manage	Process	Resolve	Work with

'the experience of generating a group report on population decline with four people helped me to understand the importance of planning, drafting and agree editing, and integrating different people's perspectives. It was tough because of the deadline, which we met, but it meant that next time we worked like that we were better at dividing the tasks and criticising and redrafting'.

Use these sites for guidance and examples of CVs. All sites accessed 16 January 2019.

Association of American Geographers, http://www.aag.org/careers

Cardiff University, Getting a job, https://www.cardiff.ac.uk/careers-advice

Careers in GIS, https://www.esri.com/en-us/what-is-gis/careers

Earthworks jobs, http://www.earthworks-jobs.com/

Environmental science: What can I do with my degree? Prospects, https://www.prospects.ac.uk/careers-advice/what-can-i-do-with-my-degree/environmental-science

Geography: What can I do with my degree? Prospects, https://www.prospects.ac.uk/careers-advice/what-can-i-do-with-my-degree/geography

Geology: What can I do with my degree? Prospects, https://www.prospects.ac.uk/careers-advice/what-can-i-do-with-my-degree/geology

Job outlook for Geoscientists, https://job-outlook.careerplanner.com/Geoscientists.cfm

Newcastle University, CVs, Applications & LinkedIn Profiles, http://www.ncl.ac.uk/careers/applications/

Prospects, CVs and cover letters, https://www.prospects.ac.uk/careers-advice/cvs-and-cover-letters

The Geological Society, https://www.geolsoc.org.uk/Geology-Career-Pathways/Careers

Use real examples all the time:

> *'In a group project (mapping/fieldwork …) I organised …, negotiated for …, shared the decision-making …'*
> *'Mentoring secondary-school students through the campus programme to create their own web pages was really rewarding. I realised I liked working both one-to-one and with small groups and that boosted my confidence to …'*
> *'I was elected treasurer of the ski club, which meant persuading people to part with subs, negotiating for union support, creating …, supporting the committee by …'*

Many university careers services have video guides to interview and assessment centre success, or search YouTube.

An application may involve you in: telephone interviews (Chapter 9.3); assessment centres; employer interviews on campus; campus careers fairs; recruitment fairs; career workshops; employer presentations; psychometric tests; standard application forms; employers' application forms and … Use the careers centre. Are there practice interview sessions? Will careers staff check your application form and CV? Bookmark the careers information websites and check regularly.

6.4　Placements, internships and volunteering

Vacation work that is paid at a good rate and provides insights into the career you think you want is exceedingly valuable, and can help significantly to boost your CV. The best advice is 'plan ahead'. These jobs are sought after and need applications up to a year in advance. You may find that you hate the company and never want to be a banker or curator or journalist, but your CV is boosted and you will not waste the first couple of years at work on something that doesn't suit you. In the best case, you may get a job for next year. Placements don't necessarily pay, but some companies and schemes offer travel expenses and a few pay very well. UK students should check out the Step scheme; it is competitive but they are very well organised, pay and have prizes. Check out **Try This 6.7** to get a feel for students' experiences and **Try This 6.8** to research volunteering opportunities.

TRY THIS　6.7 – Student placement

Here are four short reports on students' experiences of placements. What are the benefits they are reporting? What skills could they add to their CV?

Hannah Goldstein: I was thinking maybe library work, maybe museums, but I really don't know what. I contacted my local area museum, having found them on www. museumsassociation.org/home and they were really helpful. I had to send them my CV and a covering letter, saying what type of museum I was particularly interested in – there are big differences between a big interactive museum and a small traditional display case one – and how much time I could give them. In the end I opted for a placement with my local county museum, which specialises in agricultural history. I was offered four weeks during the summer. I wasn't paid, but got my travel costs and a meal allowance. I was taking phone calls from the first hour, working with visiting schools and disabled youth groups, helping with enquiries and staffing the information desk. I had a week shadowing the marketing manager, which meant I went to some interesting meetings as an observer. It really helped me to understand the business side. It was interesting work, and they asked me to come back, but it was difficult with me not earning anything while I was there.

Colette O'Connell: I had asked to work at our local newspaper and printing offices. I already had quite a lot of experience with my university newspaper on the features side, and I used to edit my school magazine, so I was expecting to contribute quite a bit. Human resources were fine about arrangements in advance, but when I turned up I was treated like a 14-year-old volunteer. I was ignored for most of the first morning. I got to do some photocopying and to help sort the post. I tried to involve myself in the news room, but they were very short-staffed and people didn't seem to have the time to help me. I explained that I had experience with newspapers, but the message didn't get through. Basically, no one knew what to do with me, and although I tried to talk to the human resources people, they were in meetings. I stopped going after three days.

Abdul Haq: I applied to the Step programme after Easter through the careers service at my university, and I was matched to a placement in my home town. I had three interviews to get my placement and it was really great. You can find out all about it on www.step.org. uk. You get paid on Step. I did ten weeks in the summer vac. They wanted a whole new section of their web pages sorted out, so I had to learn how to do that. I did the basic research to get the information I needed and to cross-check it for accuracy. I found out how to do layouts and access the corporate web design material so my new pages fitted with the company image. I had regular meetings with my supervisor, so we were both clear about the aims and my progress. She was very helpful, and checked everything before it was uploaded to the web. I was visited by somebody from the university to make sure the Step placement was working out for me and the company. Besides gaining web design and writing skills, I gained lots of office skills, and used my research skills conducting telephone interviews and a web questionnaire to get my data. My project was selected to go forward for the region to the national Step finals, so I won £500 too, as well as being paid. Great experience.

Jamie Alexander: I want to work in business, but I wasn't sure where so I applied to quite a lot of blue-chip organisations offering summer internships. It was very competitive. I had to apply online using the same form that they use for graduate recruitment, and then I had to take aptitude tests, and have an interview with a line manager. I was offered a summer internship, which was really very well paid. To a certain extent, an internship is like an extended interview: they were seeing if I would be a good employee and I wanted to know if they were right for me. I was given a specific project to do, which culminated in a report to the director of marketing, as well as going to meetings, drafting letters and dealing with clients directly. Over the summer they moved me through four departments, so I made lots of useful contacts. I realised working for an organisation of this size has lots of dimensions, and that there is great scope for promotion to different parts of the business. I went on quite a few internal training courses, and people were helpful and took time to explain what was happening to me. I applied to them in the autumn when I was back at university, and I have been offered a job to start next summer.

⚙ TRY THIS **6.8 – Researching volunteering**
What volunteering opportunities are arranged by your student union or careers centre?
What would fit around your work in term time and add to your CV?
Check out opportunities online using: *internship, volunteering, placements*.

6.5 Your own business?

Developing your own business is a career option that appeals to many students. It is the opportunity to produce and sell a product, to provide a service, or solve a problem in your community, making a difference with you in the lead. The ways in which GEES students are taught and undertake practical fieldwork develops and reinforces the skills and attributes of enterprising business people. Your degree provides communication, organisation, project planning and delivery, networking, teamworking, leadership and flexibility skills. To that list add the passion that you have to develop your product or service and make a difference in some way and you are very well equipped.

Most universities have entrepreneurship and enterprise activities in a unit/hub/ hatchery/incubator/department that helps students develop their business skills and get started. They have experts offering practical advice and skills for business planning, marketing, product design, legal and financial issues. Sessions with people running their own businesses, including students who are already running their businesses, are full of helpful insights because each business has its own issues. Entrepreneurs are usually very generous with sharing their experience, and coaching and mentoring may be available. In addition your course may include helpful modules: website development, project research and planning, economic and transport geography for example, have applications in your own business. The UK's NUS (2019) supports the sustainability and enterprise agenda with support, for example, for sustainable food–based activities on campus. Volunteering provides very valuable insights into running social enterprises and businesses.

Many businesses are started because individuals are frustrated by the lack of a service or product in their area (starting cafés, diversifying product ranges, a better web service, app development). Whatever your idea, you need to test it carefully. This is where conversations with your network of peers, in business development forums, and through the many online resources is important. A business start-up is essentially a research project. It needs an idea, researched properly to ensure there is a market, development of a business plan, clear understanding of start-up and longer-term costs, and sourcing initial sources of funding. You need to pilot, test and retest so that you know you have a large enough customer base, people who really want to pay for your product.

University enterprise and entrepreneurship hubs and incubators will let you access resources to carry out planning and research, and fully evaluate your opportunities. It is a great time to decide if this challenge will suit your career and lifestyle aspirations. Most careers centre websites have links to your university's resources, or search for: *business start-up, enterprise, social enterprise, Enactus, freelancing,* or *start-up hub.* Many companies including Dell, Facebook, Google, Reddit and Snapchat started on campus. Find out what you can do!

6.6 References and resources

British Geological Society of London 2019 Careers in the Earth Sciences, http:// www.bgs.ac.uk/vacancies/careers.htm Accessed 15 January 2019.

Enterprise Nation 2019 Student Start-up of the Year, https://www.enterprisenation. com/blog/enterprise-nation-student-start-up-of-the-year-returns-with-a-2-500-cash-prize/ Accessed 15 January 2019.

Hobson B 2018 *Why you need to embrace a portfolio career*, http://portfoliocareers. net/.

NUS 2019 Start a food enterprise, National Union of Students, https://sustainability. nus.org.uk/student-eats/start-a-food-enterprise Accessed 15 January 2019.

Prospects 2019 Country Profiles, https://www.prospects.ac.uk/jobs-and-work-experience/working-abroad Accessed 15 January 2019.

Roberts C 2010 Water Resource Scientist, https://www.youtube.com/watch?v=JFjy3lvVyuQ Accessed 15 January 2019.

Royal Geographical Society 2019 Careers with Geography, https://www.rgs.org/geography/studying-geography-and-careers/careers/finding-jobs-in-geography/ Accessed 16 January 2019.

Solem M, Foote K and Monk J 2013 *Practicing Geography: careers for enhancing society and the environment*, Association of American Geographers, Boston.

Stanley N 2019 Could a Portfolio Career be Right for you, Career Shifters, https://www.careershifters.org/expert-advice/what-to-do-when-you-want-to-do-everything-could-a-portfolio-career-be-right-for-you Accessed 15 January 2019.

The Institution of Environmental Sciences 2018 Education and Careers, https://www.the-ies.org/careers Accessed 16 January 2019.

Tieger PD and Barron-Tieger B 2014 *Do What You Are: Discovering Your Perfect Career*, (5th Edn.), Little, Brown & Company, Boston.

UCAS Teacher Training 2019 https://www.ucas.com/teaching-in-the-uk Accessed 16 January 2019.

University of Kent Careers Advisory Service 2019 I want to work in … a different way, http://www.kent.ac.uk/careers/alternatives.htm Accessed 16 January 2019.

Work2savetheworld 2011 YouTube clips, http://www.youtube.com/work2savetheworld#p/ Accessed 16 January 2019.

Volunteering (UK sites)

NCVO Directory of Volunteer Centres 2019 https://www.ncvo.org.uk/ncvo-volunteering/find-a-volunteer-centre Accessed 16 January 2019.

Projects Abroad 2019 https://www.projects-abroad.co.uk/why-projects-abroad/university-students/ Accessed 16 January 2019.

Prospects 2019 https://www.prospects.ac.uk/jobs-and-work-experience/work-experience-and-internships/volunteering Accessed 16 January 2019.

Save the Student 2019 The ultimate guide to volunteering, https://www.savethestudent.org/student-jobs/the-ultimate-guide-to-volunteering.html#abroad Accessed 16 January 2019.

Student Universe 2019 Volunteer trips, flights and accommodation, https://www.studentuniverse.co.uk/tours/volunteering Accessed 16 January 2019.

The Mighty Roar 2019 Volunteering Abroad opportunities, https://www.themightyroar.co.uk/ Accessed 16 January 2019.

Geojumble 1

Find the GEES terms in these jumbled letters. Answers in Chapter 28.

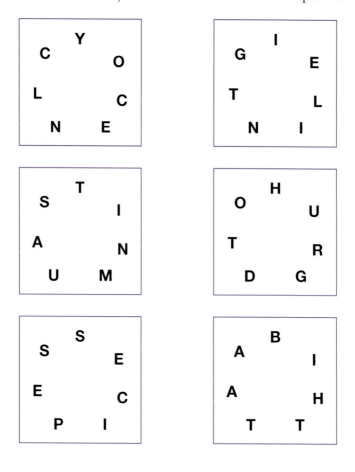

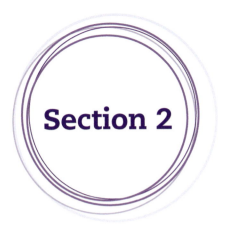

Section 2

Transforming your learning, becoming more academic!

This section spotlights different aspects of learning in higher education – with the emphasis on what is probably new or different from high school. The first chapter (7) focuses on a common problem that can knock your confidence – why some ideas and concepts are difficult to grasp – which links with Chapter 8, on thinking about how you think. This is really useful, especially if you haven't thought about how your brain works. Chapters 9 and 10 discuss listening and reading which are the core skills for learning. These are frequently overlooked because we use these skills every day without pausing to think about whether we can be more effective at learning as we listen and read. Getting the information and making sense of it needs notes (Chapter 11) so that you can build up logical arguments (Chapter 12). Becoming comfortable and confident with these academic activities requires some practice. There are lots of ways to move forward; give yourself time to experiment with different ideas and try them out.

One trick is to recognise when you are thinking on auto-pilot. For example:

What is 3 + 6?	What do you like on pizza?	What does Apple produce?

Your brain found the answers to these three questions quickly, easily. Your brain likes these types of questions. Harder questions require more thought, logical thinking, some writing and noting to pull together evidence to tell a story. For example:

? What does a low carbon economy involve?
? How are rocks classified?
? What influences sustainability?

Recognise that quick, obvious answers are probably hiding deeper, more interesting, complex ideas, and practise logical thinking.

In addition to these chapters, your university has a team of learning support staff who come to work every day to help students develop excellent learning skills and will be delighted to see you. Many of their visitors are high-performing students who want to do better, aiming to improve their assignments from high to very high marks. If you want to know more … ask them.

Relaxed maths 1

We do maths all the time. Without using a calculator solve the following:

146 ÷ 2 =	23 + 16 =	57 – 19 =
5 x 17 =	11 + 93 =	82 – 39 =
420 ÷ 5 =	81 x 7 =	35 – 12 =
45 + 6 =	13 x 15 =	224 ÷ 7 =

Answers in Chapter 28.

Threshold concepts – difficult stuff

Opening up new and previously inaccessible ways of thinking about something.

(Meyer and Land 2003)

This chapter provides a brief introduction to threshold concepts to help you to understand a little bit more about how you learn and how to learn difficult stuff. A threshold concept is something you need to understand in order to understand something else, essentially things which are important to understand so you can master the discipline. These difficult ideas are likely to be mentioned and practised in more than one module so you build up your experience and become comfortable with the idea.

University lecturers are experts at exploring what is *not* known about their discipline (the essence of research). One of the functions of a university is to explore complicated matters and 'tease out' the many dimensions and 'truths' involved through the research process. Most researchers spend their careers developing ideas and exploring their validity (research), seeing ideas evolve as new information emerges.

> **Feedback** points out good points and mistakes. Making mistakes is a normal part of learning. Learning from mistakes is important. It is rare to get something right instantly …

Lecturers are trying to sort out the confused understandings that exist in their subjects. Their challenge is to give students, usually accustomed to very structured school learning, insights into what is and what is not known, what is contested and what is agreed, and the confidence to work in the unexplored areas of the disciplines. It's brilliant when a little corner of science, social science or the humanities is made clearer. In every discipline there are difficult ideas, stuff which is hard to understand. Handling these 'threshold concepts' successfully is important for your learning.

Learning itself has thresholds. Reading is a threshold concept for early-years education. Realising that numbers are just symbols can kick-start understanding in mathematics. Think of 'passing a threshold' as a light bulb or Eureka moment in your life; it's also called transformative learning. In practice, the light bulb will flicker on and off as understanding develops: you go backwards and forwards as your understanding develops.

> *It all made sense when I listened in the lecture, and now I don't understand anything?*

How do you know when you've passed a threshold? Probably because you feel comfortable talking about the ideas in more academic or technical language. Concepts make sense in your head. It's the difference between learning Spanish, and thinking and speaking in Spanish. One benefit of developing these new understandings is seeing how different elements of a module or programme link together – you have more depth and detail.

Remember that people rarely move from not understanding to understanding in one step. Research shows that transforming your understanding, or crossing a threshold, follows on from working with lots of pieces of information in different ways: 'in short, there is no simple passage in learning from "easy" to "difficult"; mastery of a threshold concept often involves messy journeys back, forth and across conceptual terrain' (Cousin 2010: 5).

Cousin points out that the reason why lecturers seem to be so involved with their material is because they have passed a series of thresholds – have deeper understandings – which makes it hard for them to remember what it's like before things begin to 'click into place'. University gives you the time for lots of things to 'click into place', transforming your understanding. Through research you add to the 'conversations' between people that develop new ideas and understandings. You will probably transform the way that you approach learning while at university.

7.1 Scale issues

Take a moment to think about how you imagine time and space. It's big, and hard to grasp. We use models to help people understand, but models distort reality. It's important to think through where models work, and where they may generate confusion.

An orrery is a physical model of the solar system, which helps people to understand how the planets move around the Sun (look at examples online). But orreries are confusing because they are not to scale, they are *relative* representations of reality. If the Sun is modelled as a 20cm diameter ball, the Earth is the size of a peppercorn, Mars and

Deep time: Geological time, age of the planet, numbers are huge and hard to imagine.

Relative time: Subdivisions of geological time defined by the relative position of rocks, and fossil evidence.

Absolute time: Geological ages measured as accurately as currently possible, e.g. through radiometric dating of rocks and fossils.

Mercury are pinheads, and the distances between the sun and the planets is so big that you couldn't see the peppercorn Earth while standing by the model Sun. An orrery makes the planets look bigger and very much closer together than they are, which makes space travel seem more possible than it is. The diameter of the Earth and the depth of the atmosphere were very hard to understand before the first photos of the Earth were taken from the Moon. These showed how fragile and shallow the atmosphere is with respect to the diameter of the Earth. (Search: 'earth atmosphere from space' images to see this.)

The length of geological time – the time taken for tectonic plates to move, for mountains to build and evolution to take place – is difficult to imagine. The Holocene is a tiny fraction of geological history. Earth scientists talk about deep time, absolute time and relative time when constructing and reconstructing Earth history (Dodick and Orion 2006).

Geologists use the Mohs scale of hardness to compare rock mineral hardness on a relative scale. This is easier to understand once you hold and compare the different minerals. The problem with space is that you cannot stand outside the solar system and see it all, you need to use your imagination!

Understanding relationships in three dimensions (3D) is recognised as difficult for many students, which really matters when constructing cross-sections and 3D plots. In geological map work, sedimentology, paleontology and other modules, 3D practical work lets you become familiar with two- and three-dimensional graphics. Hydrologists need to understand how water flows three-dimensionally through rocks and soil to understand groundwater and run-off processes. Geomorphologists need 3D understanding to analyse landslide and glacier movements.

If you are struggling with an idea or concept, it's worth remembering that in the examples here understanding came from 'seeing'. Pictures and diagrams, sketches and doodles can be really helpful in developing understanding, rather than reading and re-reading words alone.

Geological Time Periods:

Pink **C**amels **O**ften **S**it **D**own **C**arefully, **P**erhaps **T**heir **J**oints **C**reak **P**ainfully, **E**arly **O**iling **M**ight **P**ossibly **P**revent **R**heumatism

Precambrian Cambrian Ordovician Silurian Devonian Carboniferous Permian Triassic Jurassic Cretaceous Paleocene Eocene Oligocene Miocene Pliocene Pleistocene Recent

---

7.2 Fieldwork can be transformative

GEES degrees are characterised by fieldwork. The aim is to show you what is really there. You escape the abstract notions formed through reading, and 2D lecture

> *I didn't know what the glaciology stuff was about until we went to Svalbard. Then it all made sense … You can see it all happening … and why they made us do a kit inspection first. Really cold.*

theatre slides become 3D. Understandings of geography, earth and environmental science matters happen in the field, when you get totally involved with the landscape and people. If you are not used to walking in the hills, scrambling through rivers, camping on glaciers and interviewing people, fieldwork can be difficult. Give yourself time to 'see' what is happening. Taking time to look, make notes, sketch (Chapter 27) and talk about what you see is important. Explore the area around your university, and the surrounding countryside by bus or rail. Urban or rural, they are all fantastic landscapes, well worth a 'GEES-minded' look. Taking photographs isn't a substitute, because you're not giving your brain time to really develop and process the details. Photographs can be taken very quickly and a week later it's impossible to remember why that particular picture was important (Chapters 26 and 27). Fieldwork tends to happen in separate weeks and separate modules. If cost was no problem most GEES material would be taught in the field. Sadly this is not possible. When planning a holiday, think about where would help you understand a new landscape. Got a free weekend? Explore the neighbourhood.

> *The field sketch session makes you look. I hated it because I didn't know what to draw and there is so much. It was when I looked at the tutor's sketch I began to understand … He said 'it's not art' … Add notes, and never mind the art.*

7.3 People have different thresholds

Here are reflections from students on their learning. Make some notes as you read: what are your thoughts? And consider: how might you follow up on your reflections?

> 'I suppose you could say my approach to uni involved having a great time playing hockey and then cramming at the last minute. So for me the threshold was revision, when I started doing the work. At that point you can see why different bits of the course got done, and it all began to make a bit of sense, but I don't think I really put everything together until about three months after the exams each year. Which was too late for good marks, but it did remind me why I'd wanted to do geography at uni'.

'… looking back you could say that writing an essay was a threshold for me. I did all maths and sciences at school and we didn't write an essay after I was about 13. It seemed like a junior school thing. But the lecturers just assumed I knew how to do it, so my marks weren't terribly good in first year'.

Every walk is like a field trip, even playing golf … you just see things all the time. Now I want to know what is going on.

'… the people I lived with in Hall in the first year weren't really interested in what I was. The girl who lived opposite was really brilliant at statistics and I didn't get it at all. So by Christmas I was ready to leave, thinking I couldn't pass anything. The trouble is half the modules you do have numbers in them somewhere. It wasn't like that at school. Then you get on the field class and do data, and you've got to do more stats. At least in a group there was somebody who could do them so as a group we all managed to get by. I think by the time we did the second-year field class I was understanding what was going on, but don't suppose I actually was enjoying it and it felt really unfair that other people could do it really easily'.

'… the first year skills class had that e-portfolio. Some people wrote loads and I couldn't work out what I was supposed to be writing. At the start it just didn't make sense that it will be useful to write down what we were doing. It really made sense of this last year doing the project, and for the work placement. Then it was just totally brilliant. What really changed was I got used to writing about myself, and happy about making comments about what I was doing. What was really strange was doing some writing that no one was going to mark. Why bother? And it wasn't cool. Now it makes sense but then it just seemed rubbish'.

'I think I had a real issue with understanding things I couldn't see. You just can't see radioactivity, and dating fossils and rocks and stuff. That tutorial on gravity was just the … Look, gravity, it's just there, don't mess with it. I need to be told why I need to know what gravity is, because I'm happy with it the way it is'.

'If you have never heard of stereonets, how can you be expected to understand them in one practical'.

'I did realise I needed to use a dictionary to get the words sorted out'.

*'I just didn't understand ever about how depressions work, until
I was doing a work-placement in a school and had to explain it
to my class. Now there's all these 14-year-olds who understand
depressions which is really cool, and so do I'.*

*'Numbers are okay, but what are equations? It was the second
year lab guy reading them out as words in the practicals where
it all made sense. No one really told me that the squiggly things
in the equations all mean something. I guess that sounds really
stupid but that's how it was'.*

*'We had all our first year in big lecture theatres and no one talked
to anyone. In second-year we were working in groups all the time,
and had to talk to each other. It's really hard working out that
the teacher isn't going to tell you the answer, when obviously he
knows what it is. You're used to that at school. If you wait long
enough someone will tell you the right answer. Here it is all about
working it out together, which was awesome when you got it'.*

Can you identify with any of these thoughts? Recognising that you're struggling
with understanding a particular idea can be really helpful. It should prompt you
to ask people (students, tutors, librarians, online …) whether they understand and
can explain it.

7.4 Getting unstuck

If you think you understood that first time round, you probably need to look at it again!

If you are not careful, you think about
threshold concepts as something to get over
in one step. All the evidence suggests that
people learn in cycles, so you might like to
think about how you get yourself unstuck.
Savin-Baden (2008) describes working to
get across threshold concepts as 'liquid learning'. Think of yourself as being flexible in order to deal with the real complexity of GEES materials. The point about
university is to look at complicated matters. If you think explanations are simple then you may have missed a higher-level complexity. You may think about
knowledge or information as blocks or
units, but for most issues there are different
understandings and interpretations of those
blocks of knowledge. Statistics are taught in
the first year so that they can be used in all
years. Some topics come up in a number of

It took me ages to realise that being confused at uni is normal. There are conflicting ideas about almost everything.

modules because lecturers appreciate that they are difficult, and they matter in different ways in different modules.

Different researchers have different opinions, different research evidence, and draw different conclusions. There are likely to be parallel understandings of most issues. Most GEES issues are influenced by understandings from other disciplines, for example chemistry, psychology, biology, sociology and economics, which provide further explanations and understandings of waste management, climate change, sustainability and transport systems. This diversity of understandings and ways of viewing GEES issues is referred to as contested knowledge, which is a threshold concept in its own right.

> 'Reflection is an important human activity in which people recapture their experience, think about it, mull it over and evaluate it. It is this working with experience that is important in learning'.
> (Boud et al. 1985: 19)

Learning at university is a threshold concept for most GEES students. The trick is to make time for mulling over. Enough time for liquid learning so that you 'click' at the next level. Some topics need work until your brain 'gets it'. Give yourself the time to 'Get it'.

7.5 References

Boud D, Keough R and Walker D 1985 *Reflection: turning experience into learning*, Kogan Page, London.

Cousin G 2010 Neither teacher-centred nor student-centred: threshold concepts and research partnerships, *Journal of Learning Development in Higher Education*, 2, 1–9.

Dodick J and Orion N 2006 Building an understanding of geological time, in Manduca CA and Mogk DW (Eds.) *Earth and Mind: how geologists think and learn about the earth*, The Geological Society of America, Special Paper 413, Boulder, Colorado, 77–93.

Meyer JHF and Land R 2003 Threshold Concepts and Troublesome Knowledge: Linkages to Ways of Thinking and Practising within the Disciplines, in Rust C (Ed.) *Improving Student Learning – Ten Years On*, OCSLD, Oxford, 412–424.

Savin-Baden M 2008 Liquid learning and troublesome spaces: journeys from the threshold? In Land R, Meyer RJ and Smith J (Eds.) *Threshold Concepts within the Disciplines*, Sense Publishers, Rotterdam, 75–90.

Geo-codeword 1

Substitute the numbers for letters to find the GEES-themed answers. Answers in Chapter 28.

1	2	3	4	5	6	7	8	9	10	11	12	13
				D								
14	15	16	17	18	19	20	21	22	23	24	25	26
	O						**M**					

17	5	13	7	4	11	■	17	25	17	11	17	20
8	■	■	24	■	14	■	7	■	10	■	3	
20	3	17	1	10	11	23	■	1	15	5	14	3
17	■	11	■	1	■	2	■	11	■	17	■	22
4	26	15	17	1	■	4	18	7	17	1	1	■
10	■	1	■	■	■	23	■	20	■	■	19	
15	15	1	10	11	3	■	5	3	9	11	3	20
16	■	■	20	■	17	■	■	20	■	10		
■	10	4	1	17	16	5	■	5	20	17	22	11
6	■	12	■	2	■	7	■	15	■	10	■	3
10	21	17	14	3	■	1	17	21	10	16	17	20
16	■	21	■	6	■	11	■	3	■	■	10	
19	15	2	2	3	20	■	17	4	11	26	21	17

8

Creativity, innovation and thinking

I don't think I understood what I said either.

Refusing to go out because '*I have to stay in and **think about** magma generation and crustal anomalies in the lithosphere*', or '*… the implications of positive discrimination as a factor in the self-perpetuating cycle of poverty in three developing cities*', is not cool. But creating time to think is really important. For most people, thinking starts with questions. Being asked: 'How do geothermal technologies actually work?' or 'What is your understanding of the Chicago geographers' advocacy of urban ecology as affecting our current understanding of interactions in the city?' can stimulate thoughts you didn't know you had.

> *I'd never thought about how I think, nobody ever told me it was important before. Which is really strange when you think about it. These questions seemed completely mad at first, but began to really help, reading was easier. I was taking better notes.*

Research involves generating and evaluating information, clarifying ideas, seeking new ideas, evaluating options and solving problems. Success involves thinking clearly, widely and deeply using reason, logic and imagination. Critical thinking will enable you to marshal the evidence to draw reasoned, evidenced conclusions. Creativity is core to thinking around evidence to find new ways of seeing, making links and providing new solutions to problems. Great thinking comes from being inquisitive, curious, flexible and persistent. Generally, new ideas are created following consideration of many alternatives and detailed examination from different perspectives.

Degree thinking involves knowledge from many different disciplines – economics, sociology, law, biology, chemistry, mathematics, statistics, physics, and engineering. You have the fun of extracting ideas, researching approaches and information and getting out in the field to develop and support your thinking with evidence (examples, case studies and references).

You get more intellectually sophisticated (better at thinking) during your degree course by doing stuff. This short excursion into thinking-related activities ignores most of philosophy and the cognitive sciences literature. It concentrates on unpacking what creativity, criticality and reasoning involve, and suggests questions to encourage and focus your thinking. To develop sparkling thinking, start with the ideas here. Like stunt-riding, thinking gets better with practice, not overnight. Thinking is tough, chocolate may help.

8.1 The place of creativity in thinking

Let's look at this from another angle.

University education in its broadest sense involves helping students to develop their full potential. Being creative is part of that process. Being able to explain in a job interview where you are comfortable and confident in being creative and innovative is almost certainly going to be viewed positively. It is easy to identify the analytical and practical elements in a module (see the handbook or course outline) but academics are always looking for creative elements. The aim of most academic research is to make new connections, adopt, adapt and imagine new ways of doing. Projects, essays and dissertations are usually designed so that students can develop and demonstrate their creative attributes.

Creativity involves generating new ways of thinking and doing. When a tutor asks for 'a creative solution' he is hoping you will find new ways to engage with and present your assignment. It requires practice, thinking outside of the box.

> **Creativity** – generating a new product or process.
>
> **Innovation** – developing a product or process to a new level.

Bloom's taxonomy (Figure 8.1) classifies thinking, placing creativity as the most advanced level. You already use all these thinking processes automatically in different circumstances, the trick here is to recognise the process you are using and decide if it is the most appropriate for the task; and in this way you become more conscious of your thinking processes.

	Creating. Generating new ideas, products and ways of thinking – designing, constructing, inventing, innovating , producing, planning, synthesising
	Evaluating. Justifying a decision, activity or action – experiment, check, judge, critique
	Analysing. Exploring relationships and ideas – comparing, organising, interrogating, criticising, reviewing
	Applying. Using knowledge and information in a different context – implementing, practising
	Understanding. Explaining ideas and concepts – summarising, classifying, paraphrasing, interpreting
	Remembering. Recalling information– listing, naming, finding, recognising, describing

Figure 8.1 Summary of Bloom's (1956) thinking skills taxonomy

Sternberg and Lubart (1995) take a slightly different perspective, arguing that combining three groups of abilities or attributes will lead to success in all aspects of life and learning:

- *analytical attributes* – analysis, evaluation, compare and contrast, devising a project plan …
- *practical attributes* – applying ideas, active research, fieldwork, implementing a project plan …
- *creative attributes* – imagination, synthesis, making new connections, discovering, adapting, imagining, inventing …

[**Note for later:** when you apply for a job and create your CV, how many of these thinking words can you include to describe your skills? Innovative, imaginative, curious, persistent, logical, creative, inventive, analytical, etc. See **Try This 6.5.**]

Essentially creativity involves using your imagination to develop new ideas and new ways of doing things, thinking laterally. Importantly, it is not making changes and developments for their own sake. Creative thinking leads to change where there are good reasons for change. Good ways to think about what creativity means for you are Edward de Bono's 'Six Thinking Hats', together with his other books, YouTube clips, and SCAMPER.

> Putting my ideas in the essay seemed really weird until my tutor explained how boring it is to mark 280 essays all saying what she said in the lecture.

In Six Thinking Hats activities participants take different roles. Green Hat offers new ideas, encourages brainstorming, and thinking beyond the usual solutions (De Bono 2017). In what circumstances do you use these different hats? Aim to find times to practise wearing the green hat in all parts of your life.

Creativity and innovation are terms used synonymously but which have different meanings. Creativity involves imagining and developing novel solutions. Innovation involves taking an idea or product and developing it further. A geologist imagining a completely new way to extract oil would be thinking creatively. A geologist improving an extraction technique by developing the technology is being innovative. Being creative or innovative requires action.

> **Six Thinking Hats (De Bono 2017)**
>
> White Hat asks for information.
>
> Red Hat asks about feelings, hunches and intuition.
>
> Black Hat asks difficult questions and suggests why something won't work.
>
> Yellow Hat is optimistic, cheerful; it will be okay.
>
> Green Hat looks for creative possibilities and new ideas.
>
> Blue Hat manages everyone's thinking.

SCAMPER provides a checklist to encourage you to think from different perspectives:

Substitute – seek alternatives to products, processes or roles
Combine – mix and integrate different approaches or activities
Adapt and adjust – change the expected process or procedure
Modify – modify the current process or product
Put to new uses – find new applications
Eliminate – simplify by removing unnecessary functions
Rearrange or reverse – reorder the process, activity or items

There are various websites that prompt creative approaches to research thinking; SCAMPER and Six Thinking Hats are just two good tools. Remember that enhancing creativity skills means exercising your brain. Use **Try This 8.1**. Success is making connections between concepts and ideas and presenting them with academic evidence (academic evidence means a good, reasoned and logical argument with at least two references).

⚙ TRY THIS **8.1 – Creativity and me**

These websites are good starting points. Take half an hour to explore them, then consider your approaches to problem solving, thinking around issues, and your characteristics – confidence, curiosity, ability to be persistent, for example. All websites accessed 16 January 2019.

Target Jobs 2018a Creativity: graduate recruiters like fresh thinking, https://targetjobs. co.uk/careers-advice/skills-and-competencies/300864-creativity-graduate-recruiters-like-fresh-thinking

Target Jobs 2018b Problem-solving: the mark of an independent employee, https:// targetjobs.co.uk/careers-advice/skills-and-competencies/300766-problem-solving-the-mark-of-an-independent-employee

Mind Tools 2018a Understanding Creativity, https://www.mindtools.com/pages/article/newCT_00.htm

Mind Tools 2018b SCAMPER, https://www.mindtools.com/pages/article/newCT_02.htm

Nesta 2018 Understand how innovation works, https://www.nesta.org.uk/resources

Skillsyouneed 2018 What is critical thinking? https://www.skillsyouneed.com/learn/critical-thinking.html

---

Most people agree that being creative through brainstorming, idea dumping, mind mapping, exchanging thoughts and concepts is more fun in a group. It is important to understand that people from different backgrounds, cultures and age groups will respond differently to new thoughts, concepts and products. Giving people the time to develop ideas is important. Then new ideas need testing (*evaluation*), also part of the creative process. In the GEES community, pilot projects are crucial for evaluating new ways of researching, testing, experimenting and creating new ideas.

Just Do It.

8.2 Reasoning

> I can lead my students to the reading list but I cannot make them think.

Reasoning is a term used in higher education, usually without explanation. Strong essay and examination answers look at the arguments, draw inferences and come to reasoned, logical conclusions. The arguments need to be reasoned, balanced and supported. *Supported* means *add references*.

First, consider the difference between reasoned and subjective reactions, and reflect on your thinking processes. Subjective reaction is where 'facts' are asserted without supporting evidence, whereas reasoning involves working out, or reasoning out, on the basis of evidence, a logical argument to support or disprove your case (Figure 8.2) (with at least two *references*). The subjective statement may be true, but the reader has no way of judging the quality of the information because it has no supporting evidence.

Create examples of reasoned rather than subjective statements with **Try This 8.2.** In your academic thinking and communications, avoid making emotional responses or appeals, assertions without evidence, subjective statements, analogies that are not parallel cases, and inferences based on little information, unless you qualify the argument with caveats.

Subjective statement	Reasoned statement
'Ethnic minorities are a problem'	'Smith (2030) shows that within inner city areas different ethnic groups raise different issues for social service provision. The cultural heritage and lifestyle patterns in contrasting groups means that the response to different sections within the community need to be appropriately tailored'.
'The Mississippi is a very dirty river'	'Generally the Mississippi carries a very high sediment load during flood flows because high discharges erode bed and bank material and overland flow entrains sediments from the extensive catchment (McLean and Levy 2030). The sediment load will be radically reduced immediately downstream of any of the Mississippi reservoirs due to within reservoir sedimentation from the slower moving water'.

Figure 8.2 Examples of subjective and reasoned statements

Either write a fuller, reasoned version of these five subjective statements, OR pick a few sentences from a recent essay and rewrite them with more evidence, examples and references. Sample answers are at the end of the chapter.

1. The questionnaire results are right.
2. Pelagic sediments are found in the deep oceans.
3. Man is inflicting potentially catastrophic damage on the atmosphere and causing worldwide climate change.
4. The United States has become an urban country.
5. Pedestrianisation civilises cities.

8.3 Critical thinking

Critical thinking involves working through for oneself, afresh, a problem; in that sense it is a creative process. This means starting by thinking around the nature of the problem, thinking through the issues and striving for a reasoned, logical outcome. During the process you need to be aware of other factors that impinge, where bias may be entering an argument, the evidence for and against the issues involved, and make links to other parts of your discipline and other disciplines. Earth scientists will check out the latest chemistry thinking while social geographers are reading the sociology literature. Essentially, you are critically evaluating material throughout the process of researching and thinking.

Being critical entails making judgements on information. It is important to remember that being critical does not necessarily imply being negative and derogatory; it also means being positive and supportive. It involves feeding back in a thoughtful way. A balanced critique looks at both the positive and negative aspects. Some students feel they cannot make such judgements because they

Ask a tutor how s/he assigns essay marks. Does the % balance suggested here change from Year 1 to Years 3 or 4? What is your critical content worth? Maybe:

1 **Essay writing skills 10%** – good introduction, conclusion, paragraphs for each argument, grammar and spelling.

2 **Information literacy 30%** – use of resources from the reading list 10%, and not on the reading list 20%.

3 **Critical thinking skills 60%** – logically ordered ideas, good balance of evidenced arguments and case examples, relevant material from a broad range of sources, thoughtful conclusions.

4 Anything else?

are unqualified to do so. Recognise that neither you, nor your teachers, will ever know everything: you are making a judgement based on what you know now. In a year's time, with more information and experience, your views and values may alter, but that will be a subsequent judgement made with different information.

Discussion is a major thinking aid, so talk about migration, biodiversity on coral reefs and the stratigraphy of Dorset to colleagues, friends and family. It can be provocative and stimulating! You will have new thoughts.

Where does intellectual curiosity fit into this picture? University research involves being curious about concepts and ideas. You can pursue ideas at random as they grab your imagination, but more disciplined thinking involves pursuing ideas in a logical manner, and backing up ideas and statements with solid evidence in every case.

8.4 Practising thinking

Try to avoid uncritical, surface learning, which involves listening and making notes from lectures and documents, and regurgitating the information in essays and examinations. You have acquired knowledge, but the crucial 'understanding' step is missing. Assignments which predominantly contain facts (*knowledge*) tend to get half marks. University assessments for higher marks show evidence of deeper learning, evaluation and applications or connections made to different contexts.

Adjectives used to describe **effective** thinking include: reasoned, clear, logical, precise, relevant, broad, rational, sound, sensible, critical and creative. Although daydreaming can be helpful, developing quality thinking will probably involve:

- Deciding on your objective (problem solving, understanding a concept, developing a hypothesis).
- Defining and understanding the terminology and assumptions.
- Acquiring data and information of a suitable standard to build up a reasoned argument.
- Reasoning or inferring from the available information to draw logical conclusions.
- Considering the consequences of the results, applying them back to the problem or to a new situation.

Practising this thinking sequence can take milliseconds, but when researching track the thinking progress on your tablet, notebook (lab or field notebook) so you can see ideas develop and can follow them through at different times. Ideas float away all too easily. Most issues are multidimensional (carbon capture, headwater stream ecology, geochronology, core-periphery dependency theory), which makes taking notes while you are thinking really important. Try plotting your thoughts as concept maps or spider diagrams, recording the connections and links as they occur to you. There are apps and software packages which will make your mind maps and concept diagrams look smart, but hand writing them in your notebook alongside research notes is just fine. It's the thinking and links you need – no marks for whizzy technology.

University learning should generally happen at Bloom's higher levels (Figure 8.1). You are expected to progress from remembering, which lets you acquire knowledge, to putting the emphasis on analysis, evaluation and creativity. Use **Try This 8.3** to make notes about where and how you use your brain for each of these different styles of learning.

TRY THIS 8.3 – Where and how do I use my thinking skills?	
Bloom's thinking skills	**Where and how**
Creating. Generating new ideas, products and ways of thinking – constructing, designing a process with multiple steps, developing a procedure to manage ... Assembling a team with the diverse skills to invent ...; innovation, producing, planning.	
Evaluating. Justifying a decision, activity or action – experiment, check, critique, judging which arguments or approaches are most valuable and applying them to a problem.	
Analysing. Exploring relationships and ideas – comparing, organising, interrogating, criticising, reviewing, examining the issues in turn.	
Applying. Using knowledge and information in a different context or field area – implementation, practising.	
Understanding. Explaining ideas and concepts – summarising, classifying, paraphrasing, interpreting.	
Remembering. Recalling information – listing, naming, finding, recognising, describing, repeating from notes, memorising a diagram.	

Assessment questions can prompt thinking

University assessments through essays and examination questions test the way you think about and manipulate information. You can use these exam-style questions to help you think. Questions which start with 'What do I know about …' will help to marshal information at the lower end of Bloom's scale: remembering (acquiring knowledge). This can be a good starting point BUT then move to the deeper-learning end of the scale with questions that start with:

? What criteria can be used to assess …?
? What three potential solutions you would suggest to …?
? What must be considered in order to decide about …?
? What are the most important decisions to be taken in order to …?
? How can different opinions about … be evaluated and reconciled?
? How and why would you decide to …?
? How do the different dimensions of the approaches to … change …?

The trick is to set yourself questions to answer as you research any issue. It helps focus and can change the way you make notes. Incidentally, good projects start with these types of research questions.

Questions worth asking

Judge a man by his questions rather than by his answers.
(Voltaire)

Being a critical and creative thinker involves asking questions at all stages of every research activity (lecture, tutorial, workshop, laboratory, discussion, party). The following questions could run in your head all the time. Pick a couple and ask them each day:

1. What are the main ideas here? What are the supporting ideas? What opposing evidence is available?
2. Are the questions being asked the right ones or are there more meaningful or more valid questions?
3. Is the evidence strong enough to reach a conclusion?
4. Is a particular point of view or social, scientific or cultural perspective biasing the interpretation?
5. How do these ideas fit with those read elsewhere?
6. What is assumed? Are the assumptions justified?
7. What are the strengths and weaknesses of the arguments?
8. Are the definitions right?
9. Is there evidence of high technical standards in the analysis? Is the data of an appropriate quality?
10. Do the results really support the conclusions? Are causes and effects clearly distinguished?
11. Is this information relevant? Keep thinking back to the original aims and argument. You can make statements that are clear, accurate and precise, but

if they are irrelevant they do not help. Off-the-point arguments or examples distract and confuse the reader, and may lose you marks.

12. Is this a personal opinion, an example of intuition or an evidenced statement?

13. Have I really understood the evidence – what are the main points?

14. Are these woolly, over-general statements? 'There is a serious problem with housing in America'. OK, but 'In the inner cities of the north-eastern United States there is under-provision of basic housing' (Penn and Ink 2030), or 'The extensive development of retirement housing in Florida has taken a disproportionally high percentage of the state budget for 2015 (Elder and Disney 2030)', are better.

15. Are the points made/results accurate? 'Plate margin movements are more active now than at any other time in the last 5,000 years'. Sounds clear, but is it right? Where is the evidence? Where are the references?

16. Are the results/points precise? 'The Ganges floods annually'. This statement is clear and accurate, but we do not know how precise it is. How often does the river flood? Where does it flood?

17. Is the argument superficial? Have all the complexities of an issue been addressed? 'Should slums be swept away?' A clear 'yes because …' will answer the question, but this is a very complex question, requiring consideration of the economic, social, cultural, historical, political and planning perspectives. Ask what is meant by a slum – it has different connotations in Margate and Mumbai.

18. Is there a broad range of evidence? Does the answer take into account the range of possible perspectives? The question 'Discuss the arguments for and against climate change' is so broad you cannot cover all points in equal depth. Aim to make the reader aware that you appreciate there are more points of view or approaches.

19. Are the arguments presented in a logical sequence? Check that thoughts and ideas are ordered into a sequence that tells the story in a logical and supported way.

20. How does this idea or hypothesis fit with wider fields of enquiry? You might be looking at a paper on the incidence of asthma in Indonesian communities, but where does it fit with medical geography, population geography or regional studies?

21. Which examples will reinforce the idea? What are the contradictions?

22. Can this idea be expressed in another way?

23. What has been left out? Where are the gaps?

24. Is this a definitive/true conclusion OR a probable/on-the-balance-of-evidence conclusion?

25. What are the exceptions?

Becoming an effective thinker takes time. Allow yourself to reflect before jumping into a new task with both feet. Having completed a task or activity, take a few minutes to reflect on the results or outcomes. What thought processes are being used, and what else could have been done?

Try This 8.4 and **8.5** will help to develop critical thinking skills through providing frameworks for thinking, evaluating and collating information.

TRY THIS 8.4 – Thinking about a paper

Select one paper from a reading list, any paper, any list! What Bloom (1956) level is being used? Make notes on the following:

Content: What are the main points?

Evidence: What is the support material? Is it valid? Are there references?

Counter case: What are the counter-arguments? Has the author considered the alternatives fully and referenced them?

Summary: Summarise relevant material from other sources that the author omitted but might have included.

Reflect: How well did the author meet his or her stated objectives? What research could be done next to fill in gaps or follow up on this study?

TRY THIS 8.5 – Comparing papers

Take three papers that are on the same or related topics, from any module reading list. Write (make notes for) a 1,000-word review that compares and contrasts the contributions of the three authors. (Use the guidelines from the previous exercise.) Write 250 words on where these three papers fit with material from the module.
This will take time, but it really will improve your comprehension of a topic. Treat it as a revision learning exercise. Pick three papers you are going to read anyway.

Few lecturers would argue that a logical perspective is the **only** way to deal with questions. Express your aesthetic opinions in an essay if they are relevant and appropriate. An essay on 'the nature of flow processes in rivers' requires equations for fluid flow, Reynolds numbers and case examples. A poetic answer describing the delight you feel in viewing a flowing stream at dawn will not do, and you will be in danger of being deemed 'a tributary short of a river'.

> I had a great Year 1 tutor, lovely, he talked all the time. I hated my Year 2 tutor, because she wouldn't tell us anything. But that turned out to be really useful. We had to work it out ourselves.

Can you improve the quality of your thinking alone?

Yes, but thinking takes practice, and it's easier to test your ideas in conversations. You will probably become more disciplined in your thinking by discussing issues regularly because talking around an idea sparks off other ideas in your mind. When someone else voices their point of view, you get an insight into other aspects. Thinking of arguments that run against your own position is very difficult. A discussion group might:

✓ Start by summarising the problem.
✓ Sort out objectives to follow through, a question to answer.
✓ Share data and evidence, the knowledge element.
✓ Share views on the data, *I think it means ... because ...*
✓ Work out and discuss the assumptions the data and evidence are making.
✓ Discuss possible implications; evaluate their strengths and weaknesses.
✓ Summarise the outcomes.

A good reasoner is like a good footballer, practising every day.

Thinking time

> *Did you put 'thinking time' into your plan for today? Do it now.*

Thinking happens all the time. Your brain needs time to organise information and find conclusions. A good 'discussion with thinking' conversation will have long gaps while people organise their ideas. Good tutorials have quiet tutors who ask a question, and then wait for people to offer answers, and to develop the answer in more depth. A tutorial is your thinking time, not 'teaching' time! Let people think and reply.

Where to think?

Thoughts and ideas arrive unexpectedly and drift off just as fast unless you note them. Take a minute to recall where you do your thinking. There are almost as many varied answers as people, but a non-random sample of individuals in a lecture (N=89) cited: in bed, on the train, walking to uni, running, swimming, at the gym, cleaning the kitchen and cutting grass. There is evidence of thinking being most productive while you are doing something that allows your mind to wander in all sorts of directions, without distracting phones and conversations. The majority of students citing 'walking' and 'the gym' as their best thinking opportunities demonstrate this. Capturing ideas immediately by writing or dictating to your phone, is vitally important. Take ten minutes over

> *I knew my tutorial group were winding me up by saying 'We don't know – what do you think?' But that's not the point. I do have an answer, but the group needs to learn how to get to their own answer. So I asked them to take 15 minutes to discuss, make notes and put their ideas on a poster. Then they had lots to say. It takes time for thinking to happen.*

a drink after exercise, or a couple of minutes at the bus stop, to jot down new thoughts and plans. This makes aerobic exercise an effective multitasking activity.

The history of innovation, and evidence from problem-solving activities, suggests that working with others is ideal, allowing ideas to be discussed and refined. Brainstorming is a great starter skill, letting ideas appear in an open and constructive setting.

8.5 Are you in charge of your thinking?

University students are expected to apply their already well-developed thinking skills to a series of academic tasks and activities, to make reasoned judgements and arrive at conclusions. Success involves asking questions, thinking about what you do and don't know, relating ideas and feedback from one module to other parts of your degree and to other subjects. It's all about developing your ability to interrelate evidence and draw valid conclusions. Creative thinking takes time and energy (fruit and chocolate help).

All degrees can be studied at the rather superficial level where people learn and re-present information, essentially where the three lower levels of thinking predominate (remembering, understanding, and applying). This is called surface learning and usually leads to marks in the 45%–60% range. The aim of a university education is to *really understand* through reading, discussing, practical field and laboratory work and thinking about the topics: deep learning, with the higher-order thinking skills in action (analysing, evaluating and creating).

Good thinking habits will minimise inadvertently plagiarising others' work (Chapter 15). Get into the habit of engaging and applying concepts and ideas, not just describing or reporting them. That means thinking around the ideas to find different contexts and alternative examples. Make sure you include your thoughts, opinions and reflections in your writing (Chapter 5.9). Be prepared to draft and redraft so that the thoughts are in your language, and acknowledge your sources (*add references*). Leave time to link ideas coherently. Finally, put the full reference for your citations at the end of each assignment.

Thinking involves a commentary in your head. Writing a summary in your words lets you check you understand complex ideas. Ask: '*Do I understand this?*' at the end of a page, chapter, paper, tutorial, lecture … and not just at the end. Making notes about how you are thinking, the approach you are using, is a sensible way to recognise and develop your thinking ability.

8.6 References and resources

Useful search words: *analysis, creativity, curiosity, evaluation, evidence, flexible, imagination, logic, mental model, mental map, persistent, problem-solving, reasoning.*

Adams DJ and Sparrow JC (Eds.) 2008 *Enterprise for Life Scientists*, Scion, Bloxham.

Bloom BS (Ed.) 1956 *Taxonomy of Educational Objectives: 1 Cognitive Domain*, Longman, London.

Buzan T 2009 *The Mind Map Book: Unlock your creativity, boost your memory, change your life*, BBC Books, London.

Buzan T 2010 *Use Your Head*, BBC Books, London.

Cottrell S 2011 *Critical Thinking Skills*, (2nd Edn.), Palgrave Macmillan, London.

De Bono E 2017 *Six Thinking Hats*, Penguin Books, London.

Hardy BP 2017 The 2 Mental Shifts Highly Successful People Make, https://www.thriveglobal.com/stories/4883-the-2-mental-shifts-highly-successful-people-make Accessed 15 January 2019.

Manduca CA and Mogk DW 2006 *Earth and Mind: How geologists think and learn about the Earth*, The Geological Society of America, Special Paper 413, Boulder, Colorado.

Mind Tools 2018a Understanding Creativity, https://www.mindtools.com/pages/article/newCT_00.htm Accessed 15 January 2019.

MindTools 2018b SCAMPER improving products and services, https://www.mindtools.com/pages/article/newCT_02.htm Accessed 15 January 2019.

Sternberg RJ and Lubart TI 1995 *Defying the crowd: Cultivating creativity in a culture of conformity*, Free Press, New York.

Van den Brink-Budgen R 2010 *Critical Thinking for Students*, (4th Edn.), How To Books, Oxford.

Geo-paths 2

Can you find ten or more words in the square? Move horizontally and vertically, but not diagonally, from letter to touching letter. A letter can only be used once in a single word. One word uses all the letters. Answer in Chapter 28.

E	S	T	R
R	R	T	I
R	A	X	A
E	T	E	L

Answer: Try This 8.2 – Reasoned statements

Some responses; how can you improve each one with references and more detail?

1   *The questionnaire results are right.* We can be confident of the inferences drawn from the questionnaire survey, because the sample size was large (N=678), the criteria established to ensure representative sampling were achieved, and analysis showed statistically significant relationships which confirmed the hypotheses.

2   *Pelagic sediments are found in the deep oceans.* Sediments that accumulate on the ocean floor, well away from coast lines, are dominated by clay particles and the detritus of marine organisms. These pelagic sediments are generally fine grained with clay, silica and organic oozes as the majority constituents.

3   *Man is inflicting potentially catastrophic damage on the atmosphere and causing worldwide climate change.* Information about climate change is currently speculative, but some scientific results support the hypothesis that man is damaging the atmosphere. Smoggit (2023) suggests that increasing concentrations of CO_2 and infrared absorbing gases released into the atmosphere as waste products of combustion and other human activities, are causing a net rise in global temperatures. Thermals (2030) suggests that one consequence will be higher global temperatures in the next 50 years than previously recorded.

4   *The United States has become an urban country.* At the start of the eighteenth century, 95 per cent of the population of the United States lived in rural communities or on isolated farms (Demoggers 2030). This number declined across two centuries to about 25 per cent, although of the people who live in rural areas as few as 2 per cent are employed in agriculture. The pull of industrialisation and the push of mechanisation in agriculture were major factors in this switch to an urban culture.

5   *Pedestrianisation civilises cities.* Pedestrianisation can create a safer, less polluted and apparently more spacious city. Freeing the centre of cars reduces CO and noise pollution and makes centres more attractive for pedestrians. Shoppers appreciate the calmer atmosphere, and retail therapy (shopping) is a pleasure rather than a chore (Shoppersareus 2030).

9

Listening

Hearing is easy, listening is tough.

People listen all the time, but don't always mentally process the information. GEES degrees with their mix of lectures, discussion groups and online resources can involve you in 12 or more hours of listening each week, and interviewing is a normal qualitative research methodology in geography and environmental studies, so listening is a crucial part of learning. Various studies suggest people remember less than 50 per cent of what they hear, and in a lecture the distraction of making notes can reduce this further. In addition to benefiting your academic performance, listening is increasingly cited by employers as a vital skill for business. Being a good listener, hearing and understanding most of what is said, will potentially improve your relationships with people, avoid misunderstandings, and in business you will give the impression of valuing the customer or client, taking time to understand what the speaker is saying.

Listening is not the same as hearing; it is a more active and interactive process. Listening involves being ready to absorb information, paying attention to details, and the capacity to catalogue and interpret the information. In addition to the teaching materials, there is also information in the speaker's tone of voice and body language. The more knowledge you have, the more you are likely to understand, which means that listening is a skill involving some preparation, a bit like parachuting, and it takes energy. Listening is tiring.

9.1 Listening in lectures

Arriving at a lecture with information about last night's activities or juicy scandal is normal, but the brain is not prepared for advanced information on radiolarian ooze or medieval housing. Some lecturers understand that the average student audience needs five minutes' background briefing to get the majority of brains engaged and on track. Others leap in with vital information in the first five minutes because 'everyone is fresh'! Whatever the lecturer's style, but especially with the latter, you will get

> *I am a bit dyslexic, not enough to get extra time, so being able to read the slides slowly, before the lecture, made such a brilliant difference. I was able to listen to what was said in the lecture.*

more from the session having thought in advance '*I know this will be an interesting lecture about …*' and scanned notes from the last session or library. Assuming from the start that a lecture will be dull usually ensures that it will seem dull.

> ## ⟹ TOP TIPS
>
> → Where the module PowerPoint slides or summaries are on the VLE before a lecture, look at them in advance. They won't make perfect sense, but they give you a framework for listening and you will remember much more.
> → A lecturer's words, no matter how wise, enter your short-term memory. Unless you play around with them and process the information into ideas, making personal connections, the words will drop out of short-term memory into a black hole. Think about the content and implications as the lecture progresses.
> → You may feel a lecturer is wildly off beam, making statements you disagree with, but do not decide he or she is automatically wrong – check it out. There might be dissertation possibilities.
> → Keep a record of a speaker's main points. Make notes.
> → Be prepared for the unpredictable. Some speakers indicate what they intend to cover in a lecture, others whiz off in different directions. This unpredictability can keep you alert, but if you get thoroughly lost, then ask a clarifying question (mentally or actually), rather than 'dropping out' for the rest of the session. If you feel your brain drifting off, ask questions like: '*What is s/he trying to say?*' and '*Where does this fit with what I know?*'
> → Have another look at the pages on 'Lectures' in Chapter 1.2 ('Learning opportunities').
> → Look at recorded teaching sessions within one–two weeks to clarify your notes, and decide what else to read.

9.2 Listening in discussions

No one listens until you make a mistake.

Discussion is the time to harvest ideas. With most topics in ecotourism, contaminated land management, political geography … and … there is such a diversity of points of view that open discussion is vital. Endeavour

> *I can text, talk and listen to my iPod at the same time, just don't ask me about what is going on.*

to be open-minded in looking for and evaluating statements which may express very different views and beliefs from your own. Because ideas fly around fast, make sure you note the main points and supporting evidence (arguments) where possible. Collating notes and adding references after a discussion is crucial. This involves ordering thoughts and checking arguments that support or confute the points: all the thinking skills (Chapter 8). Now look at **Try This 9.1**. How do you score on effective listening?

TRY THIS 9.1 – Self-assess your listening in discussions

Think back over a tutorial or a recent conversation. Rate and comment on your input for each of these points. Alternatively, score this for a friend, and then think about your and their listening skills.

Score your effectiveness/confidence on a 1–4 scale, where 1 is No/Low and 4 is Yes/High		Reflections/Thoughts
Did you feel relaxed and comfortable?	1 2 3 4	
Did you make eye contact with the speaker?	1 2 3 4	
Were you making notes of main points and personal thoughts during the discussion?	1 2 3 4	
Did you discuss the issues?	1 2 3 4	
Were you thinking about what to say next while the other person was still talking?	1 2 3 4	
Did you ask a question?	1 2 3 4	
Did you get a fair share of the speaking time?	1 2 3 4	
Did you empathise with the speaker?	1 2 3 4	
Did you accept what the others said without comment?	1 2 3 4	
Did you interrupt people before they had finished talking?	1 2 3 4	
Did you drift off into daydreams because you were sure you knew what the speaker was going to say?	1 2 3 4	
Did the speaker's mannerisms distract you?	1 2 3 4	
Were you distracted by what was going on around you?	1 2 3 4	

9.3 Telephone listening

Some research interviews must be conducted over the telephone, and companies regularly conduct first job interviews on the telephone or by videoconference. Listening and talking under these circumstances are difficult to do well. If Skype or video-conferencing facilities are available you can pick up on facial and body-language clues, which are missing in a telephone interview. Your telephone listening skills will improve with practice, so pilot interviews are vital: practise with a friend.

 TOP TIPS

Telephone interviews – always, every time:
☎ Find a quiet room to phone from and get rid of all distractions (radio, mobile, TV, dog …).
☎ Lay your notes out around you, have at least two pens, and if using a recorder, have it ready to start and some spare batteries.
Research interviews:

☎ Plan the call in advance. Organise your questions and comments so you can really concentrate on the responses and implications.
☎ Make notes of main points rather than every word, and leave time after the call to annotate and order the responses while the information is fresh in your brain.
☎ Query anything you are unsure about. Be certain you understand the interviewee's nuances.
☎ Show you are listening and interested without interrupting, using 'yes', 'mmm', 'OK' and 'great'.
☎ Search for verbal clues, like a changed tone of voice, to 'hear between the lines'.
☎ Don't think you know all the answers already. If you disengage, the interviewee will become less engaged, less enthused and be a less productive informant.
☎ Curb your desire to jump in and fill pauses; let the speaker do most of the talking. Silences are OK.

TOP TIPS

Job interviews on the telephone:

→ Prepare in advance as you would for a visit to a company, by researching the company background and position.
→ Make notes of points as you speak and query anything you don't understand.
→ Be enthusiastic! You have lots to offer.
→ Be formal in your conversation – there is a job in prospect. It is easy to drop into a colloquial, conversational mode as if chatting to a friend, which you would not do in an office interview.

> *I found it helped to dress like I was going to a real interview, made me concentrate … Odd at home, but no one knows.*

9.4 Listening in conversations, interviews, videos and podcasts

Remember that people can speak at approximately 125 words a minute, but you can listen and process words at 375 to 500 words a minute, so it is easy to find your brain ambling off in other directions. Wool gathering or star gazing are not good! Listening

> *I have all these notes, what did he say?*

takes concentration, and making notes is important for capturing thoughts and referring to them later. Helpfully, you can revisit information from videos and podcasts.

Jumping to conclusions while listening is dangerous as it leads you to switch off from the information. Speakers also get distracted, going in a new direction as thoughts occur to them, diverting to give additional insights. Watch out for those 'yes, but …' and 'except where …' statements.

Most of the tips for telephone interviews apply equally to personal interviews, whether for research or with an employer. If you are the interviewer, choose locations where you will not be interrupted. Take a coffee break in a long session to give both your own and your interviewee's brain a break. If you have a difficult customer, encourage them to talk. They will feel in charge and get the idea you agree with their discourse – which you may, or may not.

TOP TIPS

→ Really good listeners encourage a speaker by taking notes, nodding, smiling and looking interested and involved. Verbal feedback (replying) is often better as a statement that confirms what you have heard, rather than a question which will probably be answered by the speaker's next statement anyway ('Did you mean …?' or 'Am I right in thinking you are saying …?'). Unhelpful responses include yawning, looking out of the window, writing shopping lists and going to sleep. Relating similar personal experiences or offering solutions does not always help as, although you think you are offering empathy or sympathy, it may appear that you just turn any conversation around to yourself.

→ Remain objective and open-minded. If you are emotionally involved you tend to hear what you want to hear, not what is actually said.

→ Keep focused on what is said. Your mind does have the capacity to listen, think, write and ponder at the same time; there is time to summarise ideas and prepare questions, but it does take practice.

→ Make a real attempt to hear and understand what the other person is saying.

→ Think about what is not being said. What are the implications? Do these gaps need exploration?

→ When you are listening, interruptions need sensitive management. If you answer the phone or speak to the next person, the person you are speaking to will feel they are less important than the person who interrupts. If you do this to someone in business, they are very likely to take their business elsewhere. So turn the phone off and shut the door.

9.5 Resources

Search with *active listening*, *critical listening*, *power listening* and *listening skills* as keywords. Some of the literature relates to counselling where listening is particularly important, and Psychology has interesting and useful resources.

Grohol JM 2018 Become a Better Listener: Active Listening, https://psychcentral.com/lib/become-a-better-listener-active-listening/ Accessed 15 January 2019.

Skillsyouneed 2018 Listening Skills, https://www.skillsyouneed.com/ips/listening-skills.html Accessed 15 January 2019.

Stoker J 2016 Nine strategies for improving your listening power, https://www.careersingovernment.com/tools/gov-talk/career-advice/on-the-job/9-strategies-increasing-listening-power/ Accessed 15 January 2019.

Whitbourne SE 2018 11 Ways That Active Listening Can Help Your Relationships, https://www.psychologytoday.com/us/blog/fulfillment-any-age/201203/11-ways-active-listening-can-help-your-relationships Accessed 15 January 2019.

Geogram pairing 1

Link each word in the left-hand column to a word in the right-hand column which is its anagram. Answers in Chapter 28.

edges	shale
granite	carte
hearty	tribune
leash	inspire
point	nuclear
sample	earthy
spinier	tearing
trace	sedge
turbine	maples
unclear	piton

10

Effective reading

I read about the hazards of typhoons. I was so frightened,
I gave up reading.

Reading is a critical skill; the trick is to read and learn. Ask yourself: How do I learn when reading from paper, books, photocopies and notes, as compared with reading online from tablets, phones and computers? Online reading makes many items easily accessible. EBooks on a Kindle or computer are easy to read on trains and buses. But there is evidence that while most students say they prefer reading on screens, their actual performance tends to suffer. The evidence suggests that scrolling through pages disrupts thinking, and for longer documents it's hard to make notes and to flip forwards and back to check ideas. You can annotate some eBooks and documents, just like paper documents, but some students say they rarely look back at these notes.

The question is 'what works for you?' Maybe a good mix is the right choice: scanning online to cover options, working with printed material for detailed learning? Use this chapter to sort out what suits you in different circumstances. For a change of pace, get your computer to read a paper to you with software packages such as read&write (Texthelp 2016) – ask your university what software is available.

Reading that lets you learn takes time and patience. A good mix of paper-based and online reading helps to avoid eye-strain. Remember to slow down and enjoy reading. The brain needs space and time to absorb, understand and interact with new information.

> There were book sales in the department at the start of the year. The second-year students were really helpful, told us what was worth buying.

 TOP TIP

➜ Carry reading lists at all times.

10.1 Reading lists

Inconveniently, most reading lists are alphabetical, but you need to sort out how urgently you need to read what, and where to find the materials: online, in the library and the library location. Target more items to read than you can reasonably do in a week, so that if a book is unavailable there are alternatives.

> *I bought one book in first year, but the three of us in our flat made sure that between us we got the main books from the library, and we shared them.*

Reading lists are often dauntingly long, but you are not, usually, expected to read everything. Lists provide choice, which is especially important where class sizes are large. Serendipity cheers the brain. If a book is out on loan, don't give up. There are probably three equally good texts on the same topic on the shelf.

Also, if you find a topic confusing, reading another author's views and ideas can be very helpful. Remember that reading something is more helpful than reading nothing. RECALL books that are out of the library.

Even where there is a recommended textbook, it will rarely be followed in detail. Reading recommended books is a good idea, but watch out for those points where a lecturer disagrees with the text. Perhaps understanding of a topic has moved forward, ideas have changed. Books and journal articles take time to go through the publication process. Do you have the latest text or edition?

To get organised and decide what to read this week, make a plan:

Activity	Tasks and Issues	Key Dates
1 Sort out reading lists	Prioritise modules. Find reading lists. Underline/highlight MUST READ and MIGHT READ items.	
2 Plan reading time	Decide on what and where to read – library, online, armchair. Put times in your planner for the next three weeks. Decide what to do first on your task list.	
3 Decide what to do today	Make today's plan and get on with it!!	
4 Update task list	End of session / day. Add new papers to check out to the task list. Cross papers off the list when you are happy with your notes.	

 TOP TIP

→ Read for all modules!

10.2 Reading techniques

Printing is no substitute for reading – but it feels really, really good.

There is a mega temptation to sit down in a comfy chair with a coffee and to start reading a book at page 1. THIS IS A VERY BAD IDEA. By page 4 you will have cleaned the cat litter tray, done a house full of washing, mended a motor bike, fallen asleep or all four and more. This is great for the state of the flat, but a learning disaster. Where do you read? See **Try This 10.1**.

Everyone uses a range of reading techniques: speed-reading of novels, skip-reading headlines; the style depends on purpose. As you read this section reflect on where you use each technique already. Effective study needs 'deep study reading'.

TRY THIS 10.1 – Where and how do you read?

Here are four students' thoughts on where and how they read. Are you someone who needs pen and paper in hand to read and learn effectively? How do you answer this question?

I have to have peace to work, so we have a house rule about no music after supper for a couple of hours.

I get different opportunities to read. I am stuck with a 35-minute bus journey, which is boring but warm! I try to take scanned articles on the bus, I can't write notes but I use a highlighter pen OK-ish. I use the bus to decide if I need to make notes on something, and what just needs a five-minute entry and cross ref. in the notes I've got. For real reading I try to make notes at home, but the bus helps because I've usually got some idea about what's going on when I start.

WHERE AND HOW DO YOU READ?

The juice bar is a good place for me, not many people I know go there, so I can read, and I love the mango smoothies. I have to turn the phone off too …

My mobile has this cool app. I talk my notes and stuff to the phone and then send the doc to my email account. No one thinks it's odd, they think I am calling my mates.

Deep study reading

Deep study reading (reading when you learn as you go) is vital when you want to make connections, understand meanings, consider implications and evaluate arguments. It needs a strategic approach and time to think. Use **Try This 10.2**, where SQ3R is an acronym for Survey – Question – Read – Recall – Review. Use it NOW! The process can seem long-winded at first, but it is worth pursuing because it links thinking with reading in a flexible manner. It stops you rushing into unproductive note-making. You can use SQ3R with books and articles, and for summarising notes during revision. You are likely to recall more by using a questioning and 'mental discussion' approach to reading. Expect your reading rate to be much slower than normal. Tackling a few pages, understanding and remembering the ideas, will be more useful than covering more pages but recalling less.

TRY THIS 10.2 – SQ3R

SQ3R is a template for reading and thinking. Try it on the next book or paper you pick up.

Survey: Look at the whole book before you get into the detail. Start with the cover: is this a respected author? When was it written? Is it dated? Skim the contents page and chapter headings and subheadings to test if it will help you.

Question: Ask yourself: What do I know? What else do I want / need to know? What is new in this book that will help me? Where does this fit in this course and other modules? Decide whether deep reading and note-making is required, or whether scanning and some additions to previous notes, will suffice.

Read: Read carefully, but not necessarily from page 1. Read the sections that are relevant for you and your assignment. Read attentively but critically. On first reading, locate the main ideas. Get the general structure and subject content in your head. Then reread to make notes or highlight the essential points for your assignment.

Recall: Have a short break. Then recall (think about) what you remember, and what you understand. This process makes you an active, learning reader. Aim to explain the main points in your own words. Can you list the key points without rereading the book?

Review: Reread the text to check your understanding and clarify any points you were uncertain about. Are the headings and summaries first noted the right ones, or do they need revising? What new questions arise now that you have gone through in detail? Do you need more detail or examples?

Ask: **'Am I happy to return this book to the library?'**

Browsing

Browsing is an important research activity, used to search for information that is related and tangential to widen your knowledge. It involves giving a broader context or view of the subject, which in turn provides you with a stronger base for directed or specific reading. Browsing might involve checking out popular science, social science, and introductory texts. Good sources of general and topical geographical and environmental science information include *The Economist*, *New Scientist*, *New Internationalist* and the country and investment focus supplements in *The Guardian* and *Financial Times*. Browsing enables you to build up a sense of how the subject you are studying as a whole, or particular parts of the subject, fit together. Becoming immersed in the language and experience of the topic encourages you to think like a professional earth scientist or geographer.

Scanning

Scan when you want a specific item of information. Scan the contents page or index, letting your eyes rove around to spot key words and phrases. Chase up the references and then, carefully, read the points that are relevant for you.

Skimming

Skim-read to get a quick impression or general overview of a book or article. Look for 'signposts': chapter headings, subheadings, lists, figures; read first and last paragraphs / first and last sentences of a paragraph. Make a note of key words, phrases and points to summarise the main themes, but this is still not the same as detailed, deep reading.

 TOP TIP

➜ **Find and use examples that were not in the lecture notes.**

When reading ask yourself:

- 📖 Is this making me think?
- 📖 Am I getting a better grasp of the material?
- 📖 Am I giving myself enough time to think and read?

If the answers are no, then maybe a different reading technique would help. Reading is about being selective, and it is an iterative activity (Figure 10.1).

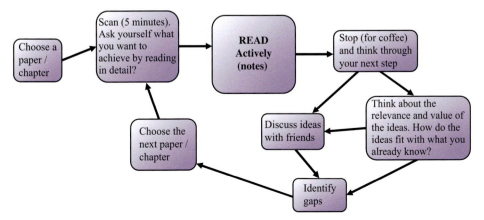

Figure 10.1 Planning reading

Building arguments

Academic journal articles and books are not thrillers. There should be a rational, logical argument, sadly rarely an exciting narrative. Usually, authors state their case and then explain the position, or argument, using careful reasoning. The writer is trying to persuade the reader (you) of the merit of the case in an unemotional and independent manner. You may feel that the writer is completely wrong, disagreeing with the case presented. If so, do not 'bin the book'; list your objections and build up your counter-arguments using other evidence. If you agree with the author, list the supporting evidence and case examples. First-class assignments report both sides of arguments with evidence (*references*). (See Chapter 12.)

Most writers use cues or signposts to guide you to important points, phrases like: 'The background indicates ...', 'the results show ...', 'to summarise ...' **Try This 10.3** will help you get comfortable with academic reading.

TRY THIS 10.3 – Cue word bingo

Find these cue words and phrases in the book / article you are reading today. How are they used? How many can you spot in a lecture? Add more phrases as you find them.

I shall outline the theory behind ...	The next point is ...
We must first examine and seek to quantify ...	Consider the issues of ...
There is little known about ...	These include ...
Recall ... Remember ...	However ...
This is the most significant advance ...	We have shown that ...
To summarise the principal points ...	We can conclude that ...
Restating the original argument in the light of this information we see ...	
We now know that ...	The important conclusion was ...

10.3 How do you know what to read?

What do I know already?

Reading and note-making will be more focused if you first consider what you already know, and use this information to decide where reading can effectively fill the gaps. Use a flow or spider diagram (Figure 10.1, Figure 12.1) to sort ideas. Put boxes around information you have already, circle areas which will benefit from more detail, check the reference list for

> *My friend, doing English, reads three books each week. How much should I be doing?*

documents to fill the gaps, and add them to the diagram. Then prioritise the circles and references, 1 to N, making sure you have an even spread of support material for the different issues. Coding notes as you read (see Chapter 11.5) and questioning encourages critical assessments, which assists in 'what to do next' decisions.

Be a critical reader

Before starting, make a list of main ideas or theories. While searching, mark the ideas that are new to you with asterisks, tick those which reinforce lecture material, and *highlight* ideas to follow up in more detail.

- Is the evidence/information from a reliable, academic source?
- Is this idea up to date?
- Are there more recent ideas?
- Do the graphs make sense?
- Are the statistics right and appropriate?
- Did the writer have a particular perspective that led to a bias in writing?
- Why did the authors research this area?
- Does their methodology influence the results in a manner that might affect the interpretation?

Library, author and journal searches start the process (Chapter 13) and practice allows you to judge the relative value of different documents. After reading, look at the author and keyword list again. Do you need to change it? Exploring diverse sources will develop your research skills. Reading and quoting sources in addition to those on the reading list may seriously impress an examiner.

- **Narrow Reading** → predictable essays and reports → middling marks
- **Wide Reading** → more creative, less predictable responses → higher marks (usually)

It doesn't usually matter what you read, or in what order. Read something.

Reading time – how long?

For most people, two hours is long enough to concentrate on one topic. A short article should take less time, but some journal articles will take longer. With longer documents you need a reading strategy, and to take breaks. Use the breaks to re-consider the SQ elements of SQ3R and decide whether your reading plan needs changing.

If you are yawning and distracted it's probably because you don't see why you should be interested. STOP READING. Use **Try This 10.2** and skim your notes to refresh your brain on WHY you are reading and what you want to get out of it. Then try again.

10.4 Styles of writing

If you find an article difficult to read, it may be because the topic is unfamiliar or the writing style is tricky. Writing styles reflect the conversational language and approach adopted in disciplines. Styles range from very direct, typically science content with information-rich short sentences, to more legal writing in environmental law and ethics modules. More philosophical literature, in cultural geography for example, is discursive. Becoming familiar with the different languages of GEES specialisms is part of being a GEES student. The *Transactions of the Institute of British Geographers* serves the entire geography community and is worth a browse for different styles of writing. *Science of the Total Environment* is always interesting and has contrasting writing styles. Earth science and geology journals have scientific styles. Think about the writing style as you read journal articles.

When you find a word you don't understand, look it up. Becoming familiar and comfortable with terminology is important for effective learning. Don't be put off by a writer's style and new words. Pick out the main message from each section. Be sure you understand the supporting and opposing arguments. Are you presenting ideas in a way that is too black and white? Are there parallel arguments? Does the text provide one side of the argument? Can you think of another side? Do you need to read something else to balance this author's view? If an article seems difficult, look at related, scene-setting materials and then reread the paper.

Think about how you read and take notes (next chapter) from factual, short-sentence styles of writing, and from more philosophical or discursive styles. The trick is to adapt your reading and note-making style to maximise your learning. This needs practice. Sharing your notes with friends can be a useful way of seeing different approaches.

 TOP TIP

✓ **Read the book once, but your notes many times.**

10.5 References and resources

Search terms: *art of reading (R), effective R, R strategies, silent or loud R, purpose of R, efficient R.*

Alexander PA and Singer LM 2017 A new study shows that students learn way more effectively from print textbooks than screen, http://www.businessinsider.com/students-learning-education-print-textbooks-screens-study-2017-10?IR= Accessed 16 January 2019.

Buzan T 2006 *Speed Reading: Accelerate your speed and understanding for success,* BBC Books, Harlow.

TCD 2017 Effective Reading, https://www.tcd.ie/Student_Counselling/student-learning/postgraduate/topics/study-skills/reading/ Accessed 16 January 2019.

Texthelp 2016 read&write, https://www.texthelp.com/en-gb/products/read-write/ Accessed 16 January 2019.

UNSW 2017 Tips for Effective Reading, https://student.unsw.edu.au/effective-reading Accessed 16 January 2019.

Add up 1

Work out the total for the top cell of each pyramid. Answers in Chapter 28.

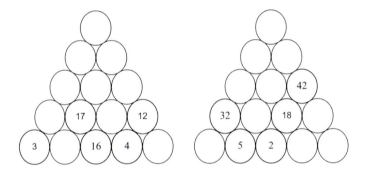

11

Making effective notes

I'm not a dictation machine, but my phone is.

Life is full of information whizzing around the net, videos, TED talks, blogs, TV reports, documentaries, lectures, tutorials, discussion groups, and in books, journals and newspapers. BUT, just because an article is in an academic journal, in the library, or on a reading list, does not make it a 'note-worthy' event. Making notes takes time: if your brain is fully involved in asking questions and commenting on the ideas, you will start to learn the material too. Noting is more than getting the facts on paper; it involves active learning and thinking to identify links

> **Note-taking or note-making?**
>
> **Note-taking** happens when everything is new. You are trying to capture facts and ideas.
>
> **Note-making** is when you look at your notes and make sense of them. You add, cross out, make links, and learn stuff.

between different pieces of information, contradictions and examples. Notes should record information in your words, evaluate different points of view, and encourage the development of your ideas and opinions. Note-making is a multi-purpose activity; like snowboarding, it gets easier with practice. Good questions to ask when making notes include: 'Is this making me think?' or 'Am I getting a clearer understanding of the topic?' Notes are usually created for private use. The major exception is field notebooks which are normally assessed for all geology field classes, and which may be assessed on geography and environmental science courses.

11.1 The note-making process

Always record your sources

Many people start reading and making notes without any sort of preview. A BAD IDEA. They make pages of notes from the opening section and few, if any, from later in the document. The first pages of a book usually set the scene. Notes are often only needed from conclusion and discussion sections. Sometimes detailed notes are required, but sometimes keywords, definitions and brief summaries are fine. Use **Try This 11.1** to evaluate how the style and length of your note-making should change for different types of information, and consider how the SQ3R method (**Try This 10.2**, p 124) fits into your note-making process. Then look at **Try This 11.2** and reflect on what you do already. What could you use in future?

TRY THIS 11.1 – Styles of note-making	
What styles of notes are needed for these different types of information? There are a couple of possible answers to kick-start your reflecting.	
Academic content – knowledge	**Style of notes**
Significant article but it repeats the content of the lecture.	None, it is in the lecture notes. BUT check your notes and diagrams are accurate. Did you note relevant sources, authors?
Fundamental background theory, partly covered in the lecture.	
An argument in favour of point x.	
An argument that contradicts the main point.	
An example from an odd situation where the general theory breaks down.	
A critically important case study.	
Just another case study.	
Interesting but off-the-point article.	A sentence at most! HOWEVER, add a cross-reference in case it might be useful elsewhere.
An unexpected insight from a different angle.	
An example/argument you agree with.	
An argument you think is unsound.	Brief notes of the alternative lines of argument, references and case examples. Comment on why it does not work so the argument makes sense to you at revision.
A superficial consideration of a big topic.	
A very detailed insight into a problem.	

⚙ TRY THIS 11.2 – How do you make notes?

Carefully consider this unordered jumble of note-making activities.
✔ those likely to assist learning, ✘ against those likely to slow up learning.
What do you do already? What could you use next time you create notes?

Leave wide margins	Code references to follow up
Identify what is not said	Store notes under the washing
Compare and revise notes with study buddies	Annotate handouts
Do loads of photocopying	Write down everything said in lectures
Copy big chunks from books	Write shopping lists in lectures
Underline main points	Ask questions
Arrange to meet friends in the library	Ignore handouts
Make notes from current affairs programmes	Jot down personal ideas
Doodle	Use coloured pens for different points
Make short notes of main points and headings	Copy all PowerPoint slides
Turn complex ideas into flow charts	Write illegibly
Ask lecturers to clarify points that make no sense	Use cards for notes
Highlight main points	Natter in lectures
Take notes from TV documentaries	Scribble extra questions in margins
Revise notes within three days of lectures	Always note references in full

11.2 Making notes from presentations

Field classes, workshops, seminars, TED talks, YouTube and lectures are awash with information, and memory meltdown syndrome will loom. Make notes of important points; you cannot hope to note everything. Listen to case studies and identify complementary examples. Highlight references mentioned by the lecturer and keep a tally of new words. Your primary goal in sessions should be to participate actively, thinking around the subject material, not to record a perfect transcript of the proceedings. Get the gist and essentials down in your own words. If sessions are recorded then you can watch or listen again.

Being critical involves questioning what you read; don't agree just because it is in print or online. Look for reasons why we should or should not accept information as true.

Constructive criticism offers suggestions to further develop or improve a piece of research or writing.

Are slides and notes on your module website? Brilliant. Reading through the slides in advance, and adding notes to them at a presentation will seriously help your learning. This is because your brain is already 'tuned in', not trying to work out what is going on. You start note-making (rather than note-taking) during the lecture.

Lecture notes made at speed in the darkness of a lecture theatre are often scrappy, illegible and usually have something missing. If you put these notes away at once you probably won't make sense of them later. Plan time to summarise and clarify notes within a day of the lecture. This reinforces ideas in your memory, hopefully stimulates further thoughts, and suggests reading priorities.

The same principles apply to creating notes from a video, YouTube clip, TED talk or podcast.

11.3 Making notes from documents

Noting from documents is easier than from lectures, because there is time to think around the issues, identify links to other material and write legibly the first time. You can read awkward passages again, but you risk writing too much. Copying whole passages postpones the hard work of thinking through the material, wasting time and paper. Summarising is a skill that develops with practice. Use **Try This 11.3** – it won't work for all articles but is a start to structuring note-making.

TRY THIS 11.3 – Creating notes from books and journals

✓ Write down the reference in full and the weblink/library location so you can find the article and journal again.
✓ Summarise the contents in two sentences.
✓ Summarise in one sentence the main conclusion.
✓ What are the strong points of the article? (Record as bullet points)
✓ Is this an argument/case I can agree with? Why?
✓ How does this information fit with my current knowledge?
✓ What else do I read to develop my understanding of this topic?

Ask yourself – where will you use the information? Scanning saves time if it avoids you making notes on an irrelevant article, or one that repeats the information you have already. In the latter case, a two-line note may be enough, e.g.,

> *'Withyoualltheway (2030) supports Originality's (2025)*
> *hypothesis with his results from a comparable study of the ecology*
> *of lemmings'*

or

> *'Dissenting arguments are presented by Dontlikeit (2025) and*
> *Notonyournellie (2026) who made independent, detailed analyses*
> *of groundwater data to define the extent of pollution from landfill*
> *leakage. Dontlikeit's main points are …'*

or

> *'Wellcushioned (2030) studying 27 retail outlets in Somerset and West Virginia, showed price fixing to be widespread. His results contrast with those of Ididitmyway's (2029) report on prices in Bangkok, because ...'*

In a report, essay or examination you have limited words and time. An essay with one case study as evidence is likely to do less well than one which covers a range of examples or cases, albeit more briefly, PROVIDED THEY ARE RELEVANT. Short notes help you to focus on arguments that are supported by a range of evidence (*references*). Make notes accordingly.

11.4 Fieldwork notes

For all GEES students accurate fieldwork notes are absolutely vital (so you can create excellent reports at another time). A small hardback notebook is ideal, with a large clear plastic bag to keep it and your hand dry as you write during monsoonal rain or a snowstorm. If you are using a tablet, phone or digital recorder, the plastic bag is crucial in the rain, and remember your spare batteries or power packs.

Notes should record your impressions and observations as accurately as possible. For geologists very accurate descriptions of the rocks, minerals, fossils and geological structures, in addition to general field sketches, are crucial. Geology tutors will explain how to create a professional-standard field notebook; it may seem tediously pernickety, but professional geologists and field environmental scientists make field observations all the time. Where your notebook is the legal record of your field observations, accuracy is very, VERY important. Field classes prepare you for this professional workplace activity through the assessment of field notebooks. Make sure all pages are dated and the location of each site is clearly marked with map or GPS coordinates. All sketches should have compass directions.

Accurate field records make writing reports much easier! Stopping to sketch and note what you see encourages looking in more depth at the landscape. Draw what is there, then add the human dimension as appropriate – traffic, types of industry, housing styles, ethnic and cultural information. Be aware of the landscape at different scales. You may be sketching one house, stream or hillside but make notes about the wider landscape, vegetation, housing and weather for context. It is not always obvious at the start what will be important for your final report. Note what you see as well as the lecturer's points.

There is a tendency to take photographs rather than make field sketches. Sketches (Chapter 27) are better because you can write notes on them, and the time it takes to complete a sketch enables you to really look and see details which are not immediately obvious. Photographs are helpful as backup but you will need to note why you took the photograph and what was important, location and aspect (GPS and compass direction). It's very hard to see the detail and make an excellent sketch at a later stage from a photo. It's the activity of drawing and adding notes to the drawing that helps you to develop the skills to understand a landscape.

11.5 Techniques

I made a mental note, but I've forgotten where I put it.

Everyone has their own approach to making notes. Aim to keep things simple, or you will take more time remembering your system than learning about ammonites, arachnids or Argentina. Try some of the following (not all at the same time).

Which medium?

Hang on while I staple this note to my tablet.

✓ Cards encourage you to condense material or use small writing. Shuffle and sort them for essays, presentations and revision.
✓ Loose-leaf paper lets you file pages at the relevant point and move pages around, which is especially useful when you find inter-module connections.
✓ Notebooks, field and lab notebooks keep everything together, but leave space as you go to add new information, feedback, comments and make links later. Index the pages so that you can find stuff!
✓ Voice recognition software/apps let you talk to the phone and upload to email or files. This can be brilliant for dictating notes when you are on the move, is fantastic for capturing ideas while running, less good in the middle of a hockey game.
✓ Using a voice recorder on your laptop, tablet or phone lets you create notes quickly. Student feedback while positive generally also cautions that '… *it's quicker, but I seem to forget what's on my phone …*'. These students suggest you need to understand how you are using different recording mechanisms and decide how you will capture information notes which become spread across different devices (paper, book, phone, laptop, iPad, tablet …).

Multicoloured highlighting

Colour-code text (online by changing the font colour or shading; handwritten with highlighter pens) to distinguish different types of information. What you highlight will depend on the task you're undertaking. Having made notes from a journal article you might, for example, highlight: key information and definitions; facts and figures worth learning; big ideas; links between things; and gaps – which may point you to further reading.

Coding

While some people like highlighting, others create a coding system to assist in dissecting structure and picking out essential points. During revision, the act of coding your notes stimulates thoughts on the types and relative importance of information. At its simplest, use a ** system in the margin:

**** Vital	** Useful
?* Possible	↓ A good idea but not for this, cross-reference to …

A more complex system can distinguish:

|| Main argument	B Background or introduction
	Secondary argument	S Summary
E.G. Case study	I Irrelevant
[Methodology, techniques	!! Brilliant, must remember
R Reservations, the 'Yes, but' thoughts	
? Not sure, need to look at/search online … to check it out.	

Opinions

Note your thoughts and opinions as you work. These are vital, BUT make sure you know which are notes from sources, feedback from the group, and which are your opinions and comments. You could use two colours, one for notes and the other for personal comments. Ask yourself questions like: 'What does this mean?' 'Is this conclusion fully justified?' 'Do I agree with the inferences drawn?' 'What has the researcher proved?' 'What is s/he guessing?' 'How do these results fit with what we knew before?' 'What are the implications for where we go next?'

Space

Leave spaces in notes, a wide margin and gaps; room to add comments and opinions later. There is no time in lectures to pursue thoughts and questions to a logical conclusion, but there is time when you are reflecting (thinking) and refocusing your thoughts and ideas.

Spider diagrams and mind maps

Turning notes into mind maps helps some people. Aim to create a visual representation of the ideas from your reading or lecture. Mind maps are especially useful for showing how ideas interrelate and the results usually help you think a little more. Search for 'spider diagram' and 'mind maps' to find out more and get software, but hand-drawn versions are what you need.

Abbreviations

Use abbreviations in notes but not essays. Txtspk may help here. Intro. = introduction; and, omitting vowels, Glc^n = glaciation, $Hist^l$ = historical, etc. Or use symbols; you probably have a system already, but here are some suggestions:

+	And	=	is the same as
→	Leads to	xxx^n	ppt^n (precipitation)
↑	Increase	xxx^g	$pump^g$ (pumping)
↓	Decrease	//	Between
>	Greater than	∴	Therefore
<	Less than	Xpt	Except
←	Before	w/	With
?	Question	w/o	Without

Millions of unordered notes will take hours to create but will not necessarily promote learning. Aim for notes that:
✓ are clear, lively and limited in length
✓ add knowledge and make connections to other material
✓ include your opinions and comments
✓ are searching and questioning
✓ guide and remind you what to do next
Finally, feeling guilty because you haven't made some, or any, notes is a waste of time and energy.

11.6 Key considerations

New words Most modules have specialist vocabulary. Keep a record of new words and check the spelling and definitions! The trick is to practise using the 'jargon of the subject' or 'geo-speak'. Get familiar with the use of words like ethnocentric, hysteresis, and pyroclastic. When you are happy with geo-terminology, you will use it effectively.

How long? '*How long should my notes be?*' is a regular student query; the answer involves lengths of string. It depends on your purpose. Generally, if notes occupy more than one side of A4, or 1–5 per cent of the original length, the topic must be of crucial importance. Some tutors will have apoplexy over the last statement: of course there are cases where notes will be longer, but aiming for brevity is good.

Sources Keep an accurate record of all your research sources, so you can cite references accurately (Chapter 14).

Quotations A direct quotation can add substance and impact, but must be timely, relevant and fully integrated in any writing. They must be fully referenced using the format: (Author date: page number), e.g. (Bloggs 2030: 432). A quote is the only time when your notes exactly copy the text.

Reproducing **maps and diagrams** is a form of quotation and again the source must be acknowledged. If you rework secondary or tertiary data in a figure, table, graph or map, acknowledge the source, as in: 'See Figure x from Mappit *et al.* (2028)'; 'The data in Table y (taken from Plottit 2031) shows that …'; or 'Re-plotting the data (Graffit 2026) shows … (Figure z)'.

Ethical reporting If you do not fully acknowledge your sources the university may impose penalties, which can range from loss of marks to dismissal from a course (Chapter 15). The regulations and penalties for plagiarism will be somewhere in your university handbook, something like: 'Any work submitted as part of any university assessment **must** be the student's own work. Any material quoted from other authors, **must** be placed in quotation marks and full reference made to the original authors'.

If you copy directly during note-making you run the risk of memorising and re-peating the material in an essay. Good note-making habits avoid plagiarism. Read and think through ideas and main points, then make notes from memory, using your words. Find new ways to express ideas. This is not as difficult as it seems, but practice helps. The original author wrote for a specific reason, but your reason and context for making notes is different. Keep asking questions and look for links to other references and modules. Notes should grab your attention, and make sense to you ten weeks later. Just swapping a couple of words is not enough. Creating diagrams, mind maps, flow or spider diagrams when making notes reduces the possibility of plagiarism. The style and language of the original author disappears behind a web of keywords and connecting arrows.

Check and share notes with study buddies. Everyone has different ideas about what is important, comparing summary notes with a mate will expand your understanding.

Be active in thinking, linking and discussing. Organising your notes, and rereading will be helpful and very useful when you are revising.

11.7 Resources

Search your university website for skills support materials. Search terms include: *visual note-taking, mind maps, doodle notes, note-making.*

Dartmouth College 2018 Notetaking, https://students.dartmouth.edu/academic-skills/learning-resources/learning-strategies/notetaking Accessed 16 January 2019.

Frank T 2014 How to Take Notes in Class: The 5 Best Methods – College Info Geek, https://www.youtube.com/watch?v=AffuwyJZTQQ Accessed 16 January 2019.

University of Nottingham 2018 Taking useful notes, https://www.nottingham.ac.uk/studyingeffectively/teaching/lecture/takingnotes.aspx Accessed 16 January 2019.

Wikihow 2018 Optimising your note-taking, https://www.wikihow.com/Take-Lecture-Notes#Optimizing_Your_Note-Taking_sub Accessed 16 January 2019.

Wordsearch 2

Locate these countries. Answer in Chapter 28.

Armenia, Belarus, Belgium, Bulgaria, Croatia, Cyprus, Denmark, Eire, Estonia, Finland, France, Germany, Greece, Italy, Lithuania, Macedonia, Malta, Monaco, Norway, Romania, Moldova, Poland, Serbia, Spain, Sweden, Turkey, Ukraine.

```
A G I T A L Y B A I M N H A L
R R O M P A U I B U R P A I E
M E K I S L N E I G O R T N S
E E Y I G O L G A L Y H L A T
N C L A D A L G A I U Y A M O
I E R E R E F N E A T K M O N
A I C U B M D I N R R A A R I
A A S A E V O I N A M U O Y A
M L A T I I A L M L K A E R S
E C N A R F O N D R A K N P C
A I B R E S E C A O R N A Y Y
A O G L U D O I A U V I D B P
S W E D E N N H T N N A P E R
C B R G N E N D A O O U B L U
L Y A W R O N Y K R L M L S S
```

12

Constructing an argument

Two geographers walking through the city observed two people shouting across the street at each other. The first said, 'Of course, they will never come to an agreement,' 'And why?' enquired his mate. 'Because they are arguing from different premises.'

At the centre of every quality GEES essay, report or presentation is a well-structured argument. Good arguments establish a set of facts, the connections that link them together, and a series of examples so that the reasoning and

> **Argument**: a fully supported and *referenced* explanation of an issue.

conclusions appear probable. You are not aiming to provide mathematical proof, or sufficient evidence to make a legal case. You are attempting to establish enough links, and to supply the case evidence through geography, earth or environmental examples, so that your argument cannot be rejected as improbable. Recognise that creating a well-organised structure, with ideas in a logical order, takes practice. There are often multiple ways to present your case, try out alternatives.

12.1 Getting organised

What is your motive for putting pen to paper (not just '*I want a 2.1*')? What do you want to say? Who is the audience? Most GEES issues can be viewed from a number of aspects, and exemplified with a wide range of material, with evidence from other disciplines (economics, physics, psychology, history …) as well. You need to unpack the elements of your argument and create a logical structure. Your aim might be to:

- develop a point of view, as in a broadsheet newspaper-style editorial or a debate;
- persuade the reader that you know about … 'post-colonial changes in agriculture in the developing world' or 'spatial modelling to forecast groundwater pollution'.

You may be more specific, perhaps wanting to define:

- how one concept differs from the previous concepts of …;
- the implications of using an idea or concept rather than another one;

- whether a decision is possible or 'Are we sitting on the fence?';
- what would need to be adapted or amended, what would work and what would not, if an idea or concept was transferred to a new situation.

> Discussions provide feedback on everyone's thoughts. Make notes. Other people's thoughts will change your ideas and develop your arguments. Appreciate all new thoughts.

Some arguments are linear, others rely on an accumulation of diverse threads of evidence which, collectively, support a particular position. Recognise that different researchers will present equally valid, but conflicting, views and opinions. In considering transport issues or mineral extraction in Devon, you could seek information from ecologists, civil engineers, agricultural scientists, transport experts, political scientists, local councillors, employment experts, environmental campaigning groups, heritage protection organisations, local people, road users and many more. In developing a rural transport statement you would present the different views and consider:

? What are the limitations of each group's view?

? What elements/ideas are entrenched?

? How might a consensus be formed?

A research project can leave you overwhelmed with evidence. 'Thinking' is crucial to designing a research approach which maximises your understanding of influences and interactions between different facets and elements of the data, and presenting these in a coherent argument. Putting main headings on a concept map, spider diagram or Post-it notes will help you visualise the different points. Post-it notes are particularly helpful because you can move them around. Personal points of view should be part of the chain in an argument, or arise from the argument. The weight of the evidence or ideas makes the point of view acceptable.

12.2 Structuring arguments

Any argument needs to be structured (Figure 12.1), with reasoned evidence supporting the statements. A stronger, more balanced argument is made when examples against the general tenet are quoted. Russell (2001) describes three classic structures that can be adopted in arguing a case (Figure 12.2). It is this author's contention that at university level, the third model should be used every time. In every discussion, oral presentation, essay, lab and field report, and dissertation you should be able to point to each of the sections and to the links between them. The key elements are the *thesis* – the argument being considered; *antithesis* – arguments against the thesis; *synthesis* – balancing various points in favour of and contradicting the argument; and *conclusion* – stating whether the thesis or antithesis is accepted or that further evidence is required.

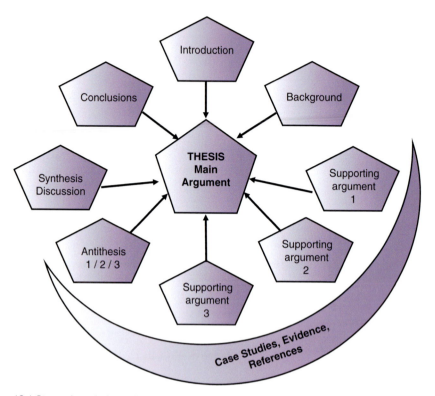

Figure 12.1 Structuring a balanced argument

1	Introduction	Thesis	Conclusion	References		
2	Introduction	Thesis	Antithesis	Conclusion	References	
3	Introduction	Thesis	Antithesis	Synthesis	Conclusion	References

Figure 12.2 Models for presenting arguments (after Russell 2001)

12.3 Unpacking arguments

It is all too easy when speaking and writing to make very general statements. One might say, 'Poverty in Africa is a consequence of the historical political systems'. This is true, but hides much information. A fuller statement like, 'Inequality in Africa has complicated historical origins. The roots of poverty in contemporary Africa can be traced back to earlier historical periods, particularly colonial times, and can be seen as a consequence of the unequal relations that exist between "North" and "South". It has also been argued that inequality in Africa is a consequence of the emergence of a capitalist world economy'.

Your level of explanation depends on context. Be careful not to explain the totally obvious. A couple of examples:

- 'Water is a liquid with chemical composition H_2O and flows downhill'. At university level we can assume people know what water is.
- 'The Richter scale measures earthquakes'. This is true but not detailed enough. 'The logarithmic Richter scale measures the energy released when the ground moves due to seismic activity. It is scaled from 0 to 10 although the largest recorded event to date is under 9.' This has additional detail and would be excellent as a starting sentence in a Level 1 essay. If you are doing a final year module on 'Earthquakes, Seismology and Management', this could be information that is already assumed to be held in common. If in doubt keep the technical content high.

Another statement meriting some unpacking is, 'Men and women play different social roles in developing countries'. A more detailed statement might say, 'Gender is an important factor in the practice and experience of cultural development. In many societies, men and women have clearly defined and often distinct roles. The different tasks assumed or assigned to men and women in each society comprise an important part of their gender identities'.

From the soils area, 'Microbiological processes control plant nutrient uptake'. This is a straightforward sentence and might be an essay title, just add 'Discuss'. A version gaining more marks by adding detail might read: 'Soil micro-organisms, which include bacteria, mycorrhizal fungi, actinomycetes, protozoa and algae, control a number of soil processes, recycling and releasing nutrients for plants and other soil organisms. The presence or absence of oxygen determines the species present and their activity rates. The mycorrhizal fungi, for example, are symbiotic dwellers on host plant tissues, taking C from the plant and making P and other nutrients available for plant uptake'.

When reviewing your own writing, first drafts of essays and reports, identify 'general' sentences which would benefit from a fuller explanation.

Q. Did you spot the major flaw in each of the examples above? What is missing?
A. There are no references, no real evidence. To strengthen each example, *add references*.

12.4 Rational and non-rational arguments

Arguments are categorised as being rational and non-rational. The non-rational are to be avoided wherever possible. Here are five examples:

- Avoid bald statements like 'Rio de Janeiro has the worst/best/smallest/ greatest … slums/geology/water supply/housing policies/landslides/ …' This is hyperbole. It might sell newspapers, but is not an argument unless supported by data.

- 'The principal problem for refugees is absence of income. Students have no income, therefore students are refugees'. This is a very poor argument and untrue in the majority of cases. Some students are refugees, most are not.
- Going OTT with language or throwing in jargon to impress the reader. This is a typical journalistic device, involving an emotional rather than factual appeal to the reader. 'Inner-city housing in Thriftston was run down, the area generally subject to neglect and depopulation' is a more considered academic statement than, 'Thriftston: an abandoned, riot-torn, rubble-strewn, neglected and deserted city', unless of course the situation is extreme. Describing Damascus in 2018 in these terms would be questionable; for central Sheffield it would be seriously biased.
- 'It is unquestionably clear that bacteria control all soil processes'. Words which sound very strong (*clearly, manifestly, undoubtedly, all, naturally* and *obviously*) will influence the reader into thinking the rest of the statement must be true. Overuse of strong words is unhelpful, the written equivalent of browbeating or shouting; used sparingly strong words have greater effect.
- Excessive generalisation: 'The Gulf of Mexico is dead' or 'All penguins are found in the Antarctic'. The global scope of geographical and environmental issues means that an exception can be found to almost every generalisation.

Inductive or deductive? Which approach?

In presenting an argument, orally or on paper, give some thought to the ordering of ideas.

A. **Inductive** writing starts with specific evidence and uses it to draw general conclusions and explanations. You must decide what suits your material. In general, use the inductive approach when you want to draw a conclusion. (Present a series of facts and case examples, and then draw your conclusions.)

B. If you write **deductively**, you begin with the general idea and then follow on with examples. The deductive approach is useful when you want to understand cause and effect, test an hypothesis or solve a problem. (Explain the general position; outline a series of arguments where A causes B, x changes y.)

An example of argument styles and good practice can be found in Holloway and Valentine (2001: 130), where the inductive method A above, is similar to their 'pulling a rabbit out of the hat' style, whereas the deductive B approach is in their 'setting out your stall' style.

Good and bad arguments

The following are rather obvious examples of illogical arguments. Be critical when reading your own drafts to spot more subtle logic problems.

☹ **Circular reasoning**: Check that conclusions are not just a restatement of your original premise. E.g. 'Urbanisation is a continuous process. We know this because we can see building going on around us all the time'.

☹ **Cause and effect**: Be certain that the cause really is driving the effect. E.g. 'Students are forced to live in crime-ridden areas, crime becomes part of the student way of life'; and 'Most faults are observed in quarries, showing that quarrying causes faulting'.

☹ **Leaping to conclusions**: The conclusion may be right, but steps in the argument are missed. Some arguments are simply wrong! For example: 'Our survey of five river reaches showed sediment grain size ranged from 0.0002 to 10mm in diameter. We therefore concluded that boulders were not a feature of this river'. Ask – did the survey select sediments of all sizes? How representative of the whole river were the five reaches?

It's important to know the difference between arguments where the supporting material provides clear, strong evidence and those where there is some statistical or experimental evidence supporting the case, but uncertainty remains. In student projects and dissertations, time usually limits experimental and fieldwork. There should be 150 samples, but, tragically, there was only time to get six. Clarity about limitations gets marks. Statements like, 'On the basis of the six samples analysed we can suggest that ...' or 'The statistical evidence suggests ... However, the inferences that may be drawn are limited because sampling occurred over one summer and at three sites in Kowloon', are very acceptable. Qualifying statements of this type have the additional merit of implying that you have thought through the limits and drawbacks inherent in your research results (*more marks*). Watch out for arguments where the author gives a true premise but the conclusion is dodgy. Just because you agree with the first part of a sentence does not mean that the second part is also right. Keep thinking right through to the end of the sentence.

Having articulated an argument, do you buy it? Why? Why not? What is your view? Look at **Try This 12.1** as a starter.

⚙ TRY THIS **12.1 – Logical arguments**

Consider the following statements. What questions do they raise? Are they true and logical? What arguments would support or refute them? There are some responses at the end of the chapter.

1. Divergent plate margins occur where two plates are moving away from each other, causing sea-floor spreading.

2. Rainforests store more carbon in their plant tissue than any other vegetation type. Burning forests release this stored carbon into the atmosphere as CO_2. The net result is increased CO_2 in the atmosphere.

3. If the system produces a net financial gain, then the management regime is successful and the development economically viable.

4. Urban management in the nineteenth century aimed to reduce chaos in the streets from paving, lighting, refuse removal and drainage, and improving law and order.

Language that persuades

How do you use linking phrases? Use **Try This 12.2** to spot those you use, and decide if there are others to add. The right link word will strengthen your argument.

TRY THIS 12.2 – Logical linking phrases

Scan through a couple of papers or books you are reading now to identify words and phrases that authors use to link arguments. Some answers are at the end of the chapter.

And keep a list of caveat or 'yes, but …' statements handy. There are alternatives to 'however', like: consequently; as a result; by contrast; thus; albeit; therefore; so; hence; nonetheless; despite the fact that; although it has been shown that. More extended versions include:

- *A rigorous qualitative geographer might argue that the results of this research are ethereal, confused and disorganised, and that some structure would have helped the project; The author gives an interesting but superficial account of …;*
- *The figures show … but were not able to support or refute the main hypothesis because …; The succeeding tests support this criticism because …;*
- *The outcome may be influenced by …;*
- *Only a few of the conclusions are substantiated by the experimental analyses.*
- *The argument is stated but the supporting evidence is not given, so we cannot …;*
- *Although this is an entirely reasonable exploratory approach, it neglects … and …, thereby weakening the inferences that may be drawn;*
- *Therefore, the criticism should be directed at … rather than at …;*
- *The outcomes, therefore, relate to a different set of conditions to those initially outlined.*

Build on this list as you read. As you research, use these 'yes, but …' phrases to focus thinking and to draw valid and reasoned inferences. Be critical (within reason) of your writing and thinking. This means allocating time to read critically, remove clichés and jargon and add caveats and additional evidence (reference, reference, reference).

Strong words often connote weak arguments.

If you use strong statements like 'Clearly we have demonstrated …', 'This essay has proved …' or 'Unquestionably the evidence has shown …', be sure that what you have written really justifies the hype.

⇨ TOP TIPS

✓ When reviewing your work, read the introduction and then the concluding paragraph. Ask: are they logically linked? Do the outcomes justify the strength of the statements?

✓ Keep your brain engaged.

12.5 Opinions and facts

It is important to make a clear distinction between facts supported by evidence, and opinions which may or may not be supported. Telling them apart takes practice. Consider (think about) the evidence offered and what else might have been said. Has the author omitted counter-arguments? Use **Try This 12.3** to help distinguish the two.

TRY THIS 12.3 – Supported facts?

Find a couple of newspapers and grab a pen.

✍ Read an article, underlining the sections that are supported facts.
✍ Identify the arguments made. Circle the opinion or commentary section.
✍ Now read 4–6 articles of different lengths and styles.
✍ Then look at the balance of opinion and facts. How do different styles of writing change the way that evidence is presented?
✍ Then play the same game with a couple of academic papers you need to read anyway. Spot how different sections are structured with evidence, and justify the author's conclusions.

12.6 Unfashionable arguments

Examining all sides of a question can be particularly difficult if moral, especially unfashionably moral or ethical elements are involved. Consider: how would you discuss the causes of urban rioting? You might say that people riot because it is fun; a group reaction; an opportunity to take personal revenge; a way of livening up a dull, hot evening; an overreaction to a minor misunderstanding that escalated beyond control and reason. You could discuss the 'fact' that some people are 'evil', 'not made evil by circumstance', and state that behaviour is not solely determined by upbringing, school, affluence and employment prospects. A more PC (politically correct) or 'right-on' answer might ignore these elements and discuss social deprivation, unemployment, police brutality, substandard housing, (mis)use of drugs and racial tensions. The rioters may be portrayed as victims of circumstance, 'it's not their fault', rather than as the active participants, inciters, throwers of bricks and bottles – people responsible for their own actions.

These issues are, of course, enormously difficult. Interlocked, interlinked and coloured in personal understanding by background and culture. The challenge for you, as an unbiased (is this possible?) reporter, is to strive to see all perspectives, including the 'moral' and the unfashionable. If the issue is 'debt', is it because 'credit' is too easily given, personally or nationally? If shoplifting is an economic problem, is it because 'shops make goods too accessible to the customer?'

One entry point might be to think of the headlines that will never make the papers:

NO RIOTING IN LOS ANGELES/LONDON/PRETORIA ON 36,518 DAYS LAST CENTURY

Why were there no riots in most cities on most nights?

 **TOP TIPS**

As you read any journal article, highlight:
→ the arguments,
→ the statements that support the argument, and
→ statements that counteract the main arguments.
Keep asking: Is the information balanced?

12.7 Examples of arguments

It is impractical in this text to reproduce an entire essay and discuss the quality of the arguments. Therefore, some students' answers, together with some examiners' feedback, are used to make points about the adequacy of the arguments. This section could be useful when revising for short-answer examination questions. Expressing ideas or concepts in two or three sentences, or a maximum of 50 words, is a good game. Consider these three answers to the question:

1. Define feedback. (Two sentences maximum.)
   A. *Feedback occurs where part or all of the output from one process is also the input for another process. This can be positive, negative or both. (27 words)*
   B. *This is where the system effectively alters itself. Positive feedback is where the signal is reinforced; the converse is called negative feedback. (23 words)*
   C. *Feedback is the action by which an output signal from a process is coupled with an input signal (Figure 12.3). Feedback may be positive, which reinforces any changes, e.g. the greenhouse effect or change in albedo, or negative, which ameliorates any changes. (49 words, including the diagram and its caption.)*

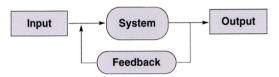

Figure 12.3 A simple feedback loop

These are not perfect answers, but the third statement has a little more technical content, examples, and the diagram reinforces the argument. This, and the next question, asks for a supported, factual response rather than an argument.

2. What is albedo?

A. *Albedo is the ratio of the total incident reflected electromagnetic energy to the total incident electromagnetic energy falling on a body. Albedo, expressed either as a % or on a 0–1 scale, varies between surfaces, and generally increases with roughness. The cryosphere has a high albedo, 70–80%, whilst the tropical rainforest has a relatively low value of 20–30%.* (59 words)

B. *Albedo = $\dfrac{\text{Total radiation reflected}}{\text{Total Incoming radiation}}$*

   It is measured on a 0–1 scale where 0 = low, 1 = high. The average albedo of the earth is 0.3. (29 words)

I think both these definitions would get full marks since both define albedo, but if there are bonus marks going, or within an essay, the first answer has added value. Using the equation format (B) perhaps makes the second answer clearer than the first. Inserting equations in essays is fine. Both answers have indicated the units involved.

Examples 3 and 4 build up an argument with evidence:

3. Explain the steps involved in deduction in science.

A. *Deduction is the logical process used where a relationship is deduced from scientific phenomena, whereas induction seeks to identify relationships within data or information. In deduction, a general idea, such as 'rivers run downhill', is formulated as a hypothesis that may be tested, as in: 'the direction of flow is determined by the gradient'. Having defined the hypothesis, it is then tested in a scientific manner in order to verify or disprove it. Tests might involve primary data collection through direct observation or analysis of maps, photographs, and records of river flow. Following analysis, the hypothesis will be accepted, rejected or accepted with caveats. Appropriate deductions might involve a statement like, 'The hypothesis was shown to be generally true for ten rivers in southern India. Further research is required to test whether these results hold true in additional areas'.*

The first sentence is a little bald (is it right?), but indicates the writer is aware of an alternative, inductive approach. The remaining sentences have integrated examples. I hope you will agree that this answer gets more marks than the following bullet-style answer:

B. *The steps in deductive reasoning are: Have an idea; develop a working hypothesis; test the hypothesis experimentally through data collection and analysis; decide whether to accept or reject the hypothesis and further explain how the world works in this case, and then reassess or redefine the problem if the research is to continue.*

This short statement answers the question at a level that will get a pass mark, but it has no discipline content linking the question to examples from geography, earth or environmental science. The first answer will score additional marks if the example used in the lecture or main text was not illustrated with river flow information.

4. Discuss the value of a structured interview compared with an online questionnaire.

Both techniques have advantages and problems. A structured interview has the benefits of being more personal, the interviewer gets a 'hands-on' feel for the data, and problems can be detected and dealt with as the interviews progress, for example by adding or dropping some questions. The interviewee should feel involved, which is likely to encourage full answers and s/he may volunteer additional information. There is a problem with the honesty of answers in all survey research, but a one-to-one interview is more likely to elicit honest responses. The number of interview refusals is likely to be small. Conversely, interviews are very time consuming in both administration and analysis, especially where there are audiotapes to interrogate.

An online questionnaire can feel very impersonal, there is little motivation to complete it (lottery/raffle rewards are unlikely to be won). Since the number of returns may be as low as 15 per cent, it must be sent to a very large population to account for non-returns. Some respondents may misunderstand questions, so instructions must be clear and questions unambiguous. However, the online questionnaire can reach a very wide national and international audience, collecting data that would be impossible face-to-face. Data analysis can be automated, and results tracked as questionnaires are completed, enabling further promotion of the questionnaire to participants in specific, and underrepresented, areas.

Both approaches require careful preparation of questions, with questions structured in a user-friendly order. A pilot survey should be carried out in each case and the schedule and analysis process adjusted as results emerge. It's important to recognise that the approach selected will generate different datasets and interpretation should identify outcomes which are an artefact of the method.

If we are looking at argument in isolation then this is good. BUT this answer would get the same marks in a sociology or psychology essay. It has argument, structure and lots of good points, but no links to geography, geology or environment, no examples and no references to the many authors discussing social survey techniques. It needs the discipline examples and understanding to do better than 45–50 per cent.

> **⇨ TOP TIPS**
>
> → Avoid 'cop-out' statements like, '*Others disagree that …*' Who are these others? There are marks for '*Knowitall (2030) and Jumptoit (2031) disagree, making the point that …*'
> → No examiner gives marks for '*etc. etc.*'

Argument Checklist

Use this when reviewing the early drafts of a report or essay. Ask: have I got …
✓ Short clear sentences, with references?
✓ A consistent deductive or inductive approach throughout?
✓ The main, supporting, and counter arguments in different paragraphs?
✓ A good balance of supporting and refuting evidence for each argument? Use Figure 12.1.
✓ Enough case examples as evidence, a balance of GEES and non-GEES examples?
✓ A high level of technical content (equations/diagrams/maps/references)?
✓ The points I want to make?
✓ The appropriate style and detail for the intended audience?

12.8 References and resources

Bonnett A 2008 *How to Argue: Essential skills for writing and arguing convincingly*, (2nd Edn.), Prentice Hall, London.

Hay I and Bass D 2002 Making News in Geography and Environmental Management, *Journal of Geography in Higher Education*, Directions, 26, 1, 129–142.

Holloway SL and Valentine G 2001 Making an Argument: writing up human geography projects, *Journal of Geography in Higher Education*, Directions, 25, 1, 127–132.

Pirie M 2015 *How to Win Every Argument: The use and abuse of logic*, (2nd Edn.), Bloomsbury Academic, London.

Russell S 2001 *Grammar, Structure and Style*, (3rd Edn.), Oxford University Press, Oxford.

University of Birmingham 2018 A short guide to essay planning and structure, https://intranet.birmingham.ac.uk/as/libraryservices/library/skills/asc/documents/public/Short-Guide-Essay-Planning.pdf Accessed 16 January 2019.

University of Bradford 2018 Developing an 'Argument' in Writing: Constructing an argument and incorporating evidence and theory, https://www.bradford.ac.uk/academic-skills/media/learnerdevelopmentunit/documents/workshopresources/howtowriteinacademicstyle/Devg-arg-CA, E, T-Booklet---Student.pdf Accessed 16 January 2019.

Geo-codeword 2

Replace the numbers with letters starting with the three indicated below. Complete the grid to find GEES-related terms. Answers in Chapter 28.

	21	16	21	11	4	13		11	21	19	9	26
1		21		4		9		18		15		9
26	21	8	4	6		15	25	4	6	26	21	2
3		15		6		12		6		26		9
11	20	3	5	21	19	9		19	15	3	5	13
4				7		5				20		
5	9	7	14	4	26		9	7	4	7	21	2
			9			24		15				21
10	21	11	7	18		4	22	3	21	11	15	6
4		9		15		11		5		6		20
26	9	17	21	6	13	20		11	6	9	16	4
20		4		20		21		6		21		7
18	4	5	1	4		19	21	8	23	26	8	

1	2	3	4	5	6	7 C	8	9	10	11 T	12	13
14	15	16	17	18	19	20 S	21	22	23	24	25	26

Answers: Try This 12.1 – Logical arguments

1. Divergent plate margins occur where two plates are moving away from each other, causing sea-floor spreading. [The first part is a reasonable definition of divergent plate margins. The second part is true where the plate margin is under the ocean. Are there divergent plate margins on land? There are examples on Iceland, and what about rift valleys? There is nothing about time scales or the rates involved.]

2. Rainforests store more carbon in their plant tissue than any other vegetation type. Burning forests release this stored carbon into the atmosphere as CO_2. The net result is increased CO_2 in the atmosphere. [Sounds OK, provided the first sentence is factually correct. If it is the *density* of the vegetation in tropical rainforests which maximises the carbon content, then the first sentence is untrue. There are other factors to consider too, like harvesting.]

3. If the system produces a net financial gain, then the management regime is successful and the development economically viable. [This is a rather sweeping statement; no time dimension is mentioned. A net profit might indicate successful management, but it might not be the managers that create the profit. What about maximising profit: could the managers do better? The development may be successful this year, but will this always be the case?]

4. Urban management in the nineteenth century aimed to reduce chaos in the streets from paving, lighting, refuse removal and drainage, and improving law and order. [The grammar here is awry. Is chaos too strong a word? Was the paving chaotic? Was there any lighting? Did urban managers have wider concerns than sorting out the streets? This is the sort of sentence where you know what the writer is trying to say, but it is slightly off-target, general and without precise examples.]

All of these statements need references inserted to provide the evidence behind the information.

Answers: Try This 12.2 – Logical linking phrases

and; if; however; do not agree; despite the fact; see; conversely; subsequently; in addition; this has been broadly related to; there may be other factors; these results are opposed to; in other words; coupled with; for example; for instance; making a causal relationship; given the pressures to; how can this; at the same time; one additional mechanism; nonetheless; such as; there is evidence; according to the theory; though varied in style; but we are not just finding; none of this is to deny; such data also; in order to understand; there are several problems with; it is more important to.

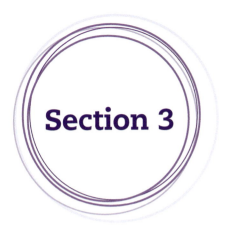

Section 3

Capturing evidence

This section focuses on skills you use to acquire the information (evidence) that characterises good-quality academic research. Success involves developing a systematic and consistent approach to research whether in the library or online (Chapter 13), collecting the information to reference your sources accurately in all assignments (Chapter 14) and behaving ethically throughout (Chapter 15). Chapter 16 looks at teamwork processes which make research and thinking so much more fun, and provide invaluable collaboration, leadership, networking and mediation experience to report when you are job hunting.

Taking a moment to reflect (*think*) on where you may be able to learn more efficiently is helpful. Where to start? What matters most? What next? On the following two sketch-graphs, plot your thoughts about tackling the next chapters. Could the diagrams be useful in helping you plan your work and other activities?

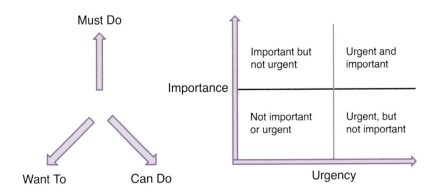

13

Research – searching for evidence

And therefore education at the University mostly worked by the age-old method of putting a lot of young people in the vicinity of a lot of books and hoping that something would pass from one to the other, while the actual young people put themselves in the vicinity of inns and taverns for exactly the same reason.

(Pratchett 1994)

Once you discover that all the notes you made so conscientiously are completely unintelligible, or you missed a lecture, using the library and online resources might be a good wheeze. Inconveniently, texts are in many places in the library. Geoscientists are likely to be heading to law, physics and engineering; geographers and environmental scientists need books and journals from agriculture, biology, chemistry, sociology, law, politics and civil engineering. This usually means books are in many locations and sometimes in different buildings. Then there are all the electronic resources. You can find eBooks and access libraries internationally. Great fun – and potentially a good way to waste time. There is a maze of information and finding the way around is not always obvious, but you CAN DO THIS. Skills you will draw on include researching, evaluation, information literacy, information retrieval, IT, flexible thinking and scheduling.

My first year was a real mess. It didn't seem cool to go to the library, it was really confusing. My tutor took us round when he realised we didn't know what to do … and then it was OK.

The mass of information available can seem scary and confusing. Most people feel very lost at first. The library catalogue is accessible online, letting you do bibliographic searches and mark up reading lists at any time. You will find online resources with a general web search, but your library has access to further resources and all the information you need. A few minutes with guided tours, watching videos and online explanations of your library's resources, getting tips on accessing library and online documents, will save you hours of time and worry. Ask library staff for help.

When researching to complete academic assignments concentrate on peer-reviewed journal papers and textbooks written by academics. The authors are academic experts providing scholarly evidence. News articles, magazines and other 'grey' literature can be equally reliable, but check the professional background of the author and decide who the item is written for – is it academically valuable? Popular articles in papers and magazines written to entertain are unlikely to provide credible academic evidence. Evaluating web resources requires the same thoughts about authorship. Websites from well-respected international organisations (United Nations, Reuters, World Wildlife Fund) and academics in universities and professional organisations can be expected to be scholarly and reliable. A comment on Twitter or Facebook doesn't have the same credibility. Deciding what is academic and scholarly is tricky, there are fuzzy boundaries. Time to exercise your critical grey matter.

In short – keep focused. Look for information which is: timely (recently published), by a credible author (an expert who is cited by others), balanced and objective (unbiased), and supported by evidence (other sources of published information).

13.1 Reading lists

Module reading lists are often long, with lots of alternatives. This is vital where classes are large. It gives you choice: each student consults a different combination of texts. Some lecturers give quite short reading lists; these may include essential reading items, and a list of authors and keywords. This approach encourages students to explore the available literature independently. Where the reading list is of the first type, it is wise to view it as a big version of the second!

Check whether your lecturer has provided references from the last two years. Some lecturers do not do so because they want to see which students have the initiative and interest to follow up on topics, to locate new research and cite it in essays, reports and exams. This allows them to credit (*give more marks to*) people who are developing as effective researchers, finding information for themselves.

How do you know whether items on a reading list are in journals, books or reports? There is a convention in citing references that distinguishes journal articles from books, and from chapters in edited books. Traditionally, a book has its TITLE in italics (or underlined in handwritten text), a journal article or report has the title of the JOURNAL in italics, and a chapter in an edited book has the BOOK TITLE in italics.

Book:

North GR and Kim K-Y 2017 *Energy Balance Climate Models*, John Wiley & Sons Ltd, Chichester.

Journal article:

Yue M, Kang C, Andres C, Qin K, Liu Y, and Meng Q 2018 Understanding the interplay between bus, metro, and cab ridership dynamics in Shenzhen, China, *Transactions in GIS*, 22, 3, 855–871.

Book chapter:

Jutagate T *et al.* 2016 Freshwater Fish Diversity in Thailand and the Challenges on its Prosperity Due to River Damming, in Orgiazzi A, Bardgett RD and Barrios E (Eds.) *Aquatic Biodiversity Conservation and Ecosystem Services*, Springer, Singapore, Chapter 3, 31–39.

In this last example you find the Jutagate *et al.* chapter, by searching for the book by Orgiazzi, Bardgett and Barrios (2016) *Aquatic Biodiversity Conservation and Ecosystem Services*.

Search for the *italicised* item first. The clue to distinguishing a journal article from a book is the numbers. In the Yue *et al.* example above, 22, 3, 855–871 indicates volume 22, issue 2, pages 855–871; books do not have this clue. Where there are no italics the game is more fun; you have to work out whether it is a journal or book you are chasing. (All students play this game; it's a university tradition.)

Before searching, highlight the papers and books on the reading list you want to read so your search is focused; look especially for Reviews and note 'linked' items to be read.

Review articles

Review articles (search topic and include the word 'review'), especially those in *Progress in Physical Geography*, *Progress in Human Geography*, *Science of the Total Environment* and *Earth Science Reviews*, provide syntheses of recent literature and point you to other references. Some journals have themed issues, or review article sections, e.g. *Environmental Science & Policy*, and *Environment and Planning A*. These articles are extended literature reviews, very helpful as preparation for project work. You may start out with one reference, but in a themed volume there will be papers on interrelated topics that are, at the least, worth browsing.

Linked reading

When reading Yue *et al.* (2018) you may realise that Yue's earlier papers, and those of other authors they cited are essential reading to get a clearer idea of the context and science. It may seem obvious, but having read article A you may need to read article B to understand A. Cross-referencing reading is a normal learning process. Remember to take accurate notes of the references that you follow up; they may prove to be the ones you cite in your writing.

 TOP TIPS

✓ Be particularly critical of sources you recognise have a bias or spin.
✓ Keep track of your reading, and a balance between searching (20 minutes) to real reading and note-making (2 hours).

13.2 Library searches

Libraries are accessed in person and from your computer. For most library visits you need a library card to get in and out, cash or card for photocopying, pen, paper, USB stick or laptop for notes. Remember to watch your bags; the opportunist thief finds a library attractive as people leave bags while searching the shelves.

At first I didn't see why you would go to the library when most of the reading was online ... I stayed in bed with the laptop to ... do stuff. But I don't get so much done at home, the library has so many more books ..., I can get more done if I am there.

Books, reports and journals

One of the changes from school to university is the emphasis on reading articles in journals. Journals contain collections of articles written by experts, published in every area of academic study. They are the way in which academics communicate their thoughts, ideas, theories and results. The considerable advantage of a journal over a book is that its publication frequency is usually six months to two years. Recent journals contain the most recent research results, which is really exciting. Reports from international and national organisations, such as the UN (United Nations), WHO (World Health Organisation), WMO (World Meteorological Organisation) etc. contain information which may not be published anywhere else. Check the location of:

✓ Recent issues of journals or periodicals held in the library. These may be stored in a different area to the older copies. Reading recent issues can give a real feel for the subject and topics of current research interest.

✓ Publications from governments and other organisations with endless amounts of vital information for GEES students.

✓ Oversized books which do not fit on the shelves are often filed at the end of a subject section, they are easy to miss – ask the librarian.

✓ Stack collections containing less commonly used books and journals, usually held somewhere else.

 TOP TIP

✓ Reserve popular books – and read them!

Unfortunately, you cannot take out all the books and keep them for the whole year. Find out what is available at other local libraries – the city or town library, for example. Many libraries offer short-loan arrangements for material that lecturers have recommended as essential. If the books you want are out, check at the shelves for similar texts that will substitute.

Return books on time. Fines can be serious, especially for restricted loans, and a real waste of good spending money. When you need a book, RECALL it. It encourages people, especially lecturers, to return them.

13.3 Electronic searches

Each university subscribes to a selection of e-resources and databases. Most are accessed with unique university passwords. You need to look at the resources online at your university library website or through Google Scholar and get used to the system. Remember that downloading eBooks, papers, abstracts, reports, and articles is not a substitute for actually reading, making notes and learning (*remembering*) the contents.

Materials available online are cited on reading lists with their URL and the date that they were accessed. This is important because web pages are updated, so readers need to know which version you are referring to:

Sunstein CR 2017 Changing Climate Change, 2009–2016, Harvard Project on Climate Agreements, Discussion Paper 2017–88, https://www.belfercenter.org/publication/changing-climate-change-2009-2016 Accessed 31 January 2019.

Many module reading lists are hyperlinked to electronic versions, making searching very fast. Remember to look at other articles in the same journal to get a broader feel for what is available – the electronic equivalent of browsing the library shelves.

For online searches use the titles, authors and keywords. There are three main Boolean operators used to refine searches: AND, OR, NOT. These operators speed up and refine a search, cutting out irrelevant sites (Figure 13.1). Entering 'Mountain AND bikes AND erosion' in Google Scholar will find 10,500 articles and patents. Refining this to articles published since 2014 cuts the list to 2,310.

Using " " (quotation marks) helps by finding the exact phrase. Searching for "mountain bikes" finds resources specifically referencing mountain bikes, ignoring resources that mention mountain and bikes individually.

Add further terms to focus your research list, reducing it to a manageable level for reading. The huge diversity of online resources means that each student finds a different combination of references and case studies, making each assignment unique.

Using 'migration OR transmigration AND gender NOT Latin America' should locate material on migration and gender issues from areas other than Latin America. Use OR when there are synonyms, and NOT to exclude topics. Using root words, as in

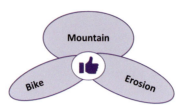

Figure 13.1 Refining searches with Boolean operators

Environ*, finds all the words that have environ as the first 7 letters: environment, environmental and environs. Beware using root* too liberally. Poli* will get politics, policy, politician, political which you might want, and policeman, polite, polish and Polish, which you might not. Remember to use synonyms (cycle, bike, bicycle, tandem) unless you want cycle, series, sequence, rotation …

It is worth an extra few minutes thinking about other terms, especially synonyms, to include in a search. American and English spellings can add confusion; use both in keyword searches. Here is a starter list, add others as you find them:

English	American	English	American	English	American
artefact	artifact	defence	defense	metre	meter
behaviour	behavior	dialogue	dialog	mould	mold
catalogue	catalog	draught	draft	plough	plow
centre	center	enclose	inclose	sulphur	sulfur
cheque	check	enquire	inquire	traveller	traveler
colour	color	foetus	fetus	tyre	tire
counsellor	counselor	labour	labor	woollen	woolen

Mineralisation (UK) and mineralization (USA) are both acceptable – pick either … ise or … ize and use consistently.

Place names: Peking and Beijing, Ceylon and Sri Lanka, Burma and Myanmar, Canton and Guangzhou. Names require careful cross-checking through atlases, with map curators or *The Statesman's Yearbook* (Turner 2018), but there are many European examples that can cause confusion, for example Lisbon and Lisboa, Cologne and Koln, Florence and Firenze.

Bookmark 'favourite' sites to save time looking for pages you use regularly. You should be able to email documents and web pages to your file space. If you open web pages and word-processing packages simultaneously to cut and paste as part of making notes be certain to mark what is quoted: **don't** plagiarise (see Chapter 15.1). When using a direct quote in your reports and essays it must be in quotation marks ('…' or "…") and properly referenced. Explore some useful sites with **Try This 13.1.** If an address is defunct, search using the site title to find the updated address, e.g. British+Geological+Survey.

⚙️ TRY THIS **13.1 – Searching online**

(all sites accessed 16 January 2019)

BBC news, http://www.bbc.co.uk/

British Geological Survey, http://www.bgs.ac.uk/

CIA World Factbook, https://www.cia.gov/library/publications/the-world-factbook/

Eldis – development policy, practice and research, http://www.eldis.org/

Environmental information for geotechnical, environmental, hydrogeology, geology, mining and petroleum topics at http://www.bgs.ac.uk/GeoIndex/home.html

Global Environmental Outlook 5, http://web.unep.org/geo/assessments/global-assessments/global-environment-outlook-5

Latin America information, http://www1.lanic.utexas.edu/

New Scientist, http://www.newscientist.com

OECD Statistics, https://data.oecd.org/

Planet Earth online, http://planetearth.nerc.ac.uk/

Reuters World News, https://uk.reuters.com/news/world

Scottish Wetlands Archaeological Database, http://xweb.geos.ed.ac.uk/~ajn/swad/introduction.html

Soilscapes, http://www.landis.org.uk/soilscapes/UK Meteorological Office, http://www.metoffice.gov.uk/

UK Office for National Statistics, https://www.ons.gov.uk/

UNDP Human Development Report, http://hdr.undp.org/en

United Nations Statistics Division, https://unstats.un.org/home/

US Census Bureau, https://www.census.gov/

World Health Organisation data, http://www.who.int/gho/database/en/

World Meteorological Organisation, https://www.wmo.int/

WARNING!

☠ Accessing databases can be totally useful or utterly frustrating. Many are accessed with an authorised login and password. Libraries cannot possibly afford to pay for access to all sites. Not being able to access a specific item will NOT cause you to fail your degree.

☠ The fact that a database exists is no guarantee that it holds the information you need.

☠ Some web documents are of limited quality, full of sloppy thinking and short of valid evidence. Some are fine. Be critical. Anyone can set up a www site. Look for reputable sites. Government and academic sites should be OK.

☠ Consider data decay carefully – which period of time does the information relate to? Economic data from the 1980s for Bosnia or the old USSR will be fine for a study of that period in those regions, but of negligible value for a current status report. Check dates carefully.

☠ Behave ethically. You **can** cut and paste from the internet to notes and essays, but all source(s) **must** be properly acknowledged. See Chapters 11 and 15.

13.4 Research strategies

A search strategy for an essay is outlined in Figure 13.2. Look at it carefully, especially the recommendation to balance time between searching and reading. Research is iterative, and helpfully online searches can be done when the library is closed. Become familiar with your local system; use **Try This 13.2** and your University library site; there will be guides to good online searching techniques for you. Good library research skills include:

✓ Using search tools such as Google Scholar efficiently.

✓ Reading and making notes.

✓ Evaluating the literature as you progress, to decide what is worth skimming, missing or reading in detail.

✓ Recording references systematically, so that referencing or continuing a search at another time is straightforward.

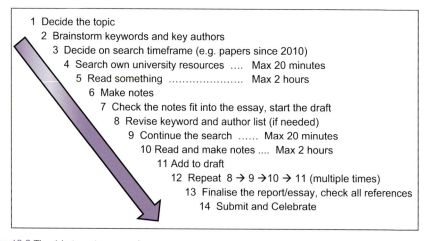

Figure 13.2 The 14 steps to research an essay

TRY THIS 13.2 – Library search
(using steps 4–6 of Figure 13.2)

Choose any topic from one of your modules. Select three authors and six keywords. Search for the papers and books in your library. Is there an interesting paper which is not on the reading list? Set a maximum of 30 minutes for online searching, then read and make notes for at least two hours.

For most undergraduate essays and projects, the resources available through your library (paper or online) are more than adequate; read these articles first. Look at wider resources later, after the first draft is written. You cannot access and read everything, essays have short deadlines and reading time is limited. The trick is to find documents available with reliable origins and at no cost.

Wikipedia

Wikipedia can be a good starting point but BEWARE. There are some very good pages. Many of the more academic entries are looked after by university lecturers keen to improve understanding of complex issues. Some sites are updated by students as course assignments, **BUT** there are many sites where the information is unreliable and biased, where you are reading someone's personal opinion or advertising spin rather than an evidenced argument. University research should always be supported by academically respectable materials (add *references*). Use Google Scholar.

> **TOP TIP**
>
> ✓ **Do not quote from Wikipedia.** Use it as a starting point. Cite the linked articles if they have real academic relevance and authority.

13.5 Ask yourself regularly – Why am I searching?

It is possible to spend all day searching online. You will acquire searching skills, know there is a paper with the ideal title in a library in Australia, or in a foreign language, and have nothing for an essay. Ignore enticing www sites initially. Searches are never done in isolation. Before starting and as you go review your reasons for researching and update your keyword searches. Put a limit on your searching time and on the types of document to include. Ideas to consider:

✓ Module essays: start with the reading list, and only explore further when you have an initial draft. Look critically at the gaps in your support material and use **Try This 13.3**.

✓ Read 'state-of-the-art-studies' and review articles written in the last 2–5 years that explore the current state of knowledge on a topic.

✓ Read an historical investigation of the development of an idea, considering how knowledge has changed over 10, 20 or more years; aim for a balance between the older and newer references.

✓ Reading a literature review should give you an outline of the 'state of the topic'. It may have a brief historical element, mapping the development of the subject knowledge, leading into a more detailed resumé of research from the past 5–10 years.

TRY THIS 13.3 – Self-assessment of a library search

Keep a tally of sources and authorities when searching. Aim to balance reading list items and your personal finds, and make sure they cover an appropriate time period.

Sources	Books	Journal Articles	Websites	Reports	Other
Articles on reading list					
Articles not on reading list					
Vintage of Sources	1800–1999	2000–2005	2006–2010	2011–2016	2017–date

---

Do not forget international dimensions: the topic might be theoretical (urban development, social housing, migration, urban climatology), but there may be regional examples that are worth considering, so check out the journals of other countries. *The Singapore Journal of Tropical Geography, Australian Journal of Earth Sciences, The Canadian Geographer, Canadian Journal of Earth Sciences, Canadian Journal of Forest Research, New Zealand Journal of Ecology, Annals of the Association of American Geography* and *Irish Geography* are all international journals, which contain articles reflecting national and regional concerns.

> WRITE UP AS YOU GO, *keep noting and drafting, and keep a record of references in full. Remember to add the reference, dates and pages on notes and photocopies.*

13.6 References and resources

Search terms: *library search, library services, library databases, information literacy.* Start with your university's information. There are excellent resources for example at the University of Melbourne (https://library.unimelb.edu.au/services/online_tutorials), and University of Southampton (http://library.soton.ac.uk/online-skills) but their specific site details are irrelevant (unless you are studying in Melbourne or Southampton).

Greenough JD and McDivitt JA 2017 Earth's evolving subcontinental lithospheric mantle: inferences from LIP continental flood basalt geochemistry, *International Journal of Earth Sciences*, 7, 1–24.

Pratchett T 1994 *Interesting Times*, Victor Gollancz, London.

Turner B (Ed.) 2018 *The Statesman's Yearbook 2018*, (16th Edn.), Palgrave Macmillan, London, http://www.statesmansyearbook.com/ Accessed 16 January 2019.

Missing vowels 2

All of the vowels are missing from these ten terms. Can you reconstruct them? Answers in Chapter 28.

RQLTY	BMSS	CHS	DLT	RSN
RRGTN	FRTLZR	GRNDWTR	ZN	RDCTV

Acknowledging sources

*Every piece of academic writing MUST include a reference list,
ALL essays, reports, practicals, posters, presentations, field
reports … ALL OF THEM.*

The previous chapter explained how to search; this one concentrates on how to cite the evidence from research in assignments. There are a number of standard ways to acknowledge research sources; journals use various styles. Most university departments have a preferred style. Look in your student handbook. LOOK NOW.

You need this chapter if there is limited or no advice in your handbook. Essentially the advice here is Harvard System with minimal punctuation to simplify your life. This system is used consistently throughout this book, providing many examples.

Top marks are awarded when every reference in the text is cited at the end in a consistent format, consistent order and punctuation. At the end of your work all references cited in the document must be listed with authors' names in **alphabetical order**, using chronological order where authors have more than one citation.

Reference list: an alphabetical record of ALL the sources cited in your document, whether you read them or not, placed at the end of the text. DO NOT include 'other things I read, but didn't quote in the text'.

Bibliography: an alphabetical list of sources or references on a particular topic; a complete bibliography would include every document ever written about a topic.

Annotated bibliography: here citations to books, journal articles and other documents are grouped together in themes. Each item or group has a brief, evaluative paragraph attached to help the reader understand more about the contents.

Your course may introduce you to EndNote (2019) or similar bibliographic software that stores reference details which you import into documents. Classes, online tutorials, YouTube, or library workshops will explain how to use the software.

14.1 Citing references in writing

When you refer to an idea, case study or quotation in any academic writing you must cite the authors. This provides the chain of evidence supporting your argument (Chapter 12). The convention is to cite the author's family name and year of publication. Cite the surname of the author and the year of publication in brackets: Moss (2026) or (Bissell and Benwell 2024), adding a, b, etc. to the year when there are two or more references to an author for the same year: (Dalrymple 2030b).

When there are two authors, both are named, but with three or more authors, the *et al.* convention is adopted. For example, 'Describing the retail geography of Great Buyit, D'Benham (2030) showed that the mega mall developments described by Markit and Hyper (2026) were already unprofitable, whereas the outlet distribution system analysed by McBurgers *et al.* (2025) had already expanded to …'. The evidence chain is completed by including these three references in full in the reference list at the end of the document.

Page numbers must always be given for quotes, conventionally using a : (Weaver 2031: 42) or Purvis (2032: 178). All direct quotes must be in 'single' or "double" quotation marks.

Please don't put full web addresses (URLs) in the document. Use the author convention, and put the URL at the end. When creating an online document with hyperlinks the URL is embedded behind the term or author of the linked connection. Watch Barley (2012) for more detail. When information on websites is updated the web address (URL) normally remains the same. Academic and government documents have unique DOI addresses.

> **URL** (Uniform Resource Locator) – the web address
>
> **DOI** (Digital Object Identifier) – an internationally standardised system for uniquely locating documents

> PLEASE use *et al.* correctly.
>
> Really nice academics are annoyed when *et al.* is misused. It must always be in italics because it is Latin meaning 'and other persons', with a full stop after *al.* because *al.* is an abbreviation of *alia*.

You are citing references to improve the quality of your argument by providing evidence, which sometimes involves multiple documents where previous authors have cited other people. Where information in one text refers to another, quote both: 'Stokes (2026, cited in Law 2031) found that …'. Both the Stokes (2026) and Law (2031) references must appear in the reference list. Similarly: 'In an extensive review of kluftkarren, Eastwood (2034) shows the field approach taken by

Clint (2029) is unreliable, and therefore the methodology adopted by Clint is not followed'. Quote both sources, although you have probably only read Eastwood. Quoting both tells the reader how to locate the original. To make clear that you have acquired your information from a secondary source, use a sentence like 'Landslides in North America annually injure 5,000 people and cause property damage in excess of $12 billion (Crushum 2028, cited in Flattenem *et al.* 2031)'. In this case, it is important to give both dates to indicate the age of the original data, 2021, rather than the 2032 date of the reference you read. You must put both Crushum and Flattenem *et al.* in your references. The Crushum reference will be cited in the Flattenem *et al.* paper, so not including it is lazy. If Flattenem *et al.* does not list Crushum, use your library or other search engines to find it.

In some publications the author is not immediately obvious. The author may be a Committee, as in many House of Commons publications, or an organisation, or the publisher. There are no absolute rules; use common sense or follow the pattern in journal reference lists.

Referencing by using abbreviations or initials can be convenient and time saving, but may cause confusion. Wherever they are used, the full name must be provided somewhere in the reference. Cite HoC EAC (2017) in your text, but ensure the initials are explained within the reference list:

HoC EAC 2017 *Environmental Impact of Microplastics*, Fourth Report of Session 2016–17, HC 179, House of Commons Environmental Audit Committee, The Stationery Office, Norwich.

An Act of Parliament is cited by its title and date in your text, for example '… in the Neighbourhood Planning Act (2017) …'. In the reference list it is:

Neighbourhood Planning Act 2017 *Public General Acts*, The National Archives, http://www.legislation.gov.uk/ukpga/2017/20/contents Accessed 15 January 2019.

If there doesn't seem to be a rule on abbreviations, and you cannot find a precedent, invent one and use it consistently.

Be careful when inserting copied or scanned diagrams. They need the reference in the usual way with the **page number for the diagram** included. For example, when copying Figure 4 from Boy and Uitermark (2017) you must include in your citation the figure and page number from the original paper: Boy and Uitermark (2017, Fig. 4: 621). The reference list cites the paper in full:

Boy JD and Uitermark J 2017 Reassembling the city through Instagram, *Transactions of the Institute of British Geographers*, 42, 612–624, https://onlinelibrary.wiley.com/doi/epdf/10.1111/tran.12185 Accessed 15 January 2019.

Where a publication is only online the page numbers will be omitted.

14.2 Citing paper-based sources in your reference lists

The key is consistency in format: use a standard sequence of commas, stops, spaces and italics. Underline the italicised items in handwritten documents. The reason italics are used is because they tell you where to locate the document within edited and multiple-author publications.

Citing a book – The *book title is in italics*: Author(s) Year *Title*, (Edition), Publisher, Place of publication.

Lamb D 2017 *Large Scale Forest Restoration*, (2nd Edn.), Routledge, London.

Woodcock NH and Strachan R 2012 *Geological History of Britain and Ireland*, (2nd Edn.), Wiley-Blackwell, Oxford.

Ballas D, Clarke G, Franklin R and Newing A 2018 *GIS and the Social Sciences*, Routledge, London.

Where there are multiple editions of the book, it's important to state which edition you have read, as contents and page numbers change with updating in the newer editions. The convention is (2nd Edn.) (3rd Edn.) … The Edn. element is not needed for first editions.

It can be difficult to identify the place of publication where international publishers list multiple cities. Either use the first city listed or where the publisher's headquarters is located.

Occasionally, an organisation or publisher is the author, as for example:

Collins Maps 2016 *Collins World Atlas*, (12th Edn.), Collins, London.

Citing an edited book – Editors' names (Eds.) Year *Title*, (Edition), Publisher, Place of publication.

Koch R and Latham A (Eds.) 2017 *Key Thinkers on Cities*, Sage, London.

Citing a chapter in an edited book – The authors of the chapter or paper in an edited volume are cited first, followed by the book editors' details. Only the *title of the book is italicised*, not the chapter title: Author(s) Year Chapter title, in Editors' names (Eds.) *Volume title*, Publisher, Place of publication, (optional – Chapter number), Page numbers.

Brantut N and Platt JD 2017 Dynamic weakening and the depth dependence of earthquake faulting, in Thomas MY, Mitchell TM and Bhat HS (Eds.) *Fault Zone Dynamic Processes: evolution of fault properties during seismic rupture*, John Wiley & Sons, Chichester, Chapter 9, 171–194.

Citing a journal article – Here the *Journal title is italicised*, and the URL or DOI added if the article is online. The second example below doesn't include page numbers because at the time it was viewed for this publication it was only available online.

Author Year Article title, *Journal title*, Volume number, Issue number, Page numbers, DOI/URL Accessed date.

Vögeli N, van der Beek P, Huyghe P and Najman Y 2017 Weathering in the Himalaya, an East-West comparison: indications from major elements and clay mineralogy, *The Journal of Geology*, 125, 5, 515–529, https://www.journals.uchicago.edu/doi/abs/10.1086/692652 Accessed 15 January 2019.

Woods N, Fowler P and Siu S 2018 Stability assessment of the Anderson Road Quarry face Hong Kong, *Quarterly Journal of Engineering Geology and Hydrogeology*, 23 March, https://doi.org/10.1144/qjegh2017-121 Accessed 15 January 2019.

In the rare case where there is no author attribution, use the Anon convention:

Anon 2021, Rebuilding Ank-Morpork, *Discworld Journal of Architecture*, 23 May, 4–5.

Citing a newspaper article – When an author is cited use: Author Full date Title, *Newspaper*, Volume number if applicable, Page number(s). DOI/URL Accessed date.

Hern A 5 April 2018 Facebook among 30 organisations in UK facing political data inquiry, *The Guardian*, https://www.theguardian.com/technology/2018/apr/05/facebook-mark-zuckerberg-refuses-to-step-down-or-fire-staff-over-mistakes Accessed 15 January 2019.

When there is no author cited, use the first words of the headline in your text citation: 'Organic Farm (2022)'; and put the full headline in the reference list:

Article title, Full date, Newspaper, Volume number if applicable, Page numbers.

Organic Farm Revolutionizes Veg Deliveries, 16 April 2032, *The Borchester Echo*, 5.

Citing an unpublished thesis – Thesis citations follow the general guidelines for a book, but add 'unpublished', the degree title awarded, and enough information for another researcher to locate the volume. If available online add the DOI or URL. Traditionally, italics are not used, indicating this is an unpublished item.

Valenci M 2017 Heavy metal contamination of river water: sources, behaviour and remediation, unpublished PhD thesis, University of Leeds, http://etheses.whiterose.ac.uk/18772/ Accessed 15 January 2019.

Sandy IAM 2021 Coastal management influences on the beach morphology of the Norfolk Coast, unpublished BSc dissertation, School of Geography, Earth and Environmental Science, University of Poppleton.

14.3 Citing e-sources

NEVER, EVER put a web address in an essay, report …
The URL ALWAYS goes in the reference list, at the end.

The formats for citing e-sources, films, TV and other materials are settling into a pattern. These notes follow recommendations from various online and library sources. When writing for a publication check and follow their style guide. The crucial element is adding the date when you accessed the website, because the contents of sites change. Some Wikipedia entries change daily. The next person to access the site may not see the same information. Did I say – **Never put the full** http://www… addresses within your essay or report text, only in reference lists?

Citing web pages – Treat internet and other electronic sources like paper-based references in text. For example '… the estimated and projected HIV/AIDs-related deaths for China (UNAIDS 2018) indicate that …'.
　　Web addresses tend to be long, so they need careful checking. The pattern is:
　　Author/Editor Year Title (Edition), Publisher, Place of publication, URL Accessed date.

UN 2018 Careers: job openings, United Nations, New York, https://careers. un.org/lbw/home.aspx?viewtype=SJ&exp=INT&level=0&location=All&occup=0&department=All&bydate=0&occnet=0 Accessed 31 January 2019.

Publication dates are not always obvious, so check at the foot/end of the web page sometimes there is a copyright date. If it's really, really missing, write 'no date'. The 'Accessed date' shows when you viewed the document. 'Publisher' covers both the traditional idea of a publisher of printed sources, and organisations responsible for maintaining sites on the internet. Many internet sites show the organisation maintaining the information, but not the text author. If in doubt, ascribe authorship to the smallest identifiable organisational unit.

Citing online journals – The citation process is the same as described above for hard copy journals. Either the DOI or URL is required for locating an article.
　　Author Year Title, *Journal title*, Volume, Issue, Page numbers, DOI/URL Accessed date.

Citing email – To reference personal email messages, use the 'subject line' of the message as a title and include the full date. Remember to keep copies of the emails you reference.

Sender (Sender's email address), Full date of email, Subject of message, email to Recipient (Recipient's email address).

Solo H (hans.solo@carida.ac.cal), 21 July 2032, Essay for second tutorial, email to Chewbacca, W. (Chewbacca.wookie@carida.ac.cal).

Citing blogs – Author Year posted Title of blog post, Blog name, Date of posting, URL Accessed date.

Altaweel M 2018 Geography and religion, GeoLounge, 3 April, https://www. geolounge.com/geography-and-religion/ Accessed 15 January 2019.

Citing wikis – Wikis can be very awkward to cite because there are multiple authors, and sites on main platforms such as Wikipedia are often updated. For example, the Camelford water pollution incident in 1988 has had updates since Wikipedia began and these continue. The View History link provides the date of the last revision. This page shows 231 revisions and 81 editors. By providing the Accessed date, another reader can track and view the version that you read; it may have been changed more recently.

Author Year of last revision Title of entry, Wiki name, URL Accessed date.

Anon 2018 Camelford water pollution incident, Wikipedia, http://en.wikipedia. org/wiki/Camelford_water_pollution_incident Accessed 31 January 2019.

Citing social networking websites – I strongly advise you use such references very rarely in academic assignments. They are not usually strong academic evidence. Where necessary use:

Author Year Message title, Network site, Date, URL Accessed date.

Simon C 2031 Answer for fieldtrip group work, Facebook, 12 January, http:// en-gb.facebook.com/1234567890 Accessed 17 February 2031.

Citing Twitter – This follows the format for a social network site.
Author Year Title of Tweet, Platform, Date, URL Accessed date.

Watson J 1887 Holmes suggests scarlet for study, Twitter, 14 September, http:// twitter.com/WatsonJohn/221b Accessed 17 February 1888.

Citing a podcast or video – Author Year Title, Platform, URL Accessed date.

Thompson R 2015 Plastic waste in the ocean, YouTube, https://www.youtube. com/watch?v=ZGNEuivjBak Accessed 15 January 2019.

If in doubt use a short version of the title in your essay or a sensible shorthand. There is a Nelson Mandela University Geology Honours Field Trip video on You-Tube. This could be referred to as NMU (2017), or NMU Geology (2017). The reference gives the full details:

NMU 2017 Geology Honors & 3rd Year Trip to Plettenberg Bay and Humansdorp, Nelson Mandela University, YouTube, https://www.youtube.com/watch?v= zvCwlaCjPgc Accessed 31 January 2019.

Citing a film – in the text of an essay use the title, director and date. For example: '… in Blade Runner 2049 (dir:Villeneuve 2017), we see …' and in the reference list:

Villeneuve, D (dir) 2017, Blade Runner 2049, Sony Pictures, USA.

Take time to double-check the details for any web posting to make sure you are citing as accurately as possible. Authors' names and publisher information is often hidden at the foot of pages, or in links embedded in the picture, film or credits!

⇨ TOP TIPS

✓ The author's name may be found at the foot of an electronic document. Where the author is unclear, the URL should indicate the name of the institution responsible for the document. However, this organisation may only be maintaining the document, not producing it. Take care to assign the right authorship.
✓ Date of publication. This is often at the foot of the page with the author's name, and sometimes with 'last updated' information.
✓ Keep accurate records of the material you access, notes, spread sheet …

14.4 Citing maps

Maps may be hard copy or electronic. Citing **the scale** is important in all map references.

A **map extract** in the text or Figure title is cited as 'extract from Isle of Man map' (Ordnance Survey, 2009), and in the references needs:

Map maker Year of issue Title of map, Map series, Sheet number, Scale, Publisher, Place of publication, Page number(s).

Ordnance Survey 2009 Isle of Man, Landranger Series, Sheet 95, 1:50000, Ordnance Survey, Southampton.

An atlas is cited in the same way as a book, and a map within a book needs its title and page number as in the second example:

Dorling D and Thomas B 2016 *People and Places: a 21st century atlas of the UK*, Policy Press, Bristol.

Brooke-Hitching E 2016 Australia's Inland Sea, Scale not given, in Brooke-Hitching E *The Phantom Atlas: the greatest myths, lies and blunders on maps*, Simon & Schuster, London, 34.

14.5 Citing photographs

All photographs, including those taken by you, must acknowledge the photographer's name with the image, and the full reference at the end. This may include: Year of production, Title of image, and where it is located, for example a library or gallery location. (See Figure 22.2, p 263). For personal photos use:

Figure xx. Cliffs at Charmouth, 8.30 am, 15 June 2018, Dorset. Photograph by the author.

And in references:

Davis, J. 2018 Personal photograph, Jura field class, Scotland. 30 September.

Attributions can lead to lengthy Figure titles. For a neat example see Figure 1 in:

Upton P, Koons PO and Roy SG 2018 Rock failure and erosion of a fault damage zone as a function of rock properties: Alpine Fault at Waikukupa River, New Zealand, *Journal of Geology and Geophysics*, DOI: 10.1080/00288306.2018.1430592 Accessed 16 January 2019.

> 'Figure 1. **A**, <u>Google Earth</u> image of the study area including the Waikukupa River and Hare Mare Creek. The active Hare Mare fault trace and the abandoned Waikukupa thrust are shown with linking strike-slip segments. Fault traces and lithologic boundaries <u>modified from Norris and Cooper (1995, 1997)</u>. The yellow star marks the approximate location of the photo shown in C. **B**, Location diagram showing the South Island, the Alpine Fault and relative plate vector <u>from DeMets et al. (1994)</u>. **C**, Example of gullying from the abandoned Waikukupa strand of the Alpine Fault'.

In this Figure legend and consistently in their paper the authors (Upton *et al.* 2018) used underlining to highlight their acknowledgements.

14.6 Citation practice

Producing correct reference lists is an important skill, demonstrating your attention to detail and professionalism. Spot the errors in **Try This 14.1**. The ultimate test of a reference list is that someone else can use it to locate the documents. Check your citations meet this standard.

⚙️ TRY THIS **14.1 – Spot the referencing errors**

How many can you find? Answers are at the end of the chapter.

1 Clapham A 2018 Human Rights: a Very Short Introduction, (Second Edition). Oxford

2 World Economic and Social Survey 2017 by United Nations

3 Wright RT and Boorse DF 2017 Environmental Science: Toward a Sustainable Future, 13th Edn., Pearson,

4 Clapham 2015 Sediment Transport, https://www.youtube.com/watch?v=kKXd0dv6ae4

5 **Korkiamäki R and Kallio** KP 2018 Experiencing and practising inclusion through friendships, Area, 74–82

6 MacDonald Magee and Goodenough K.M. 2017 Dykes as physical buffers to metamorphic overprinting: an example from the Archaean–Palaeoproterozoic Lewisian Gneiss Complex of NW Scotland, Scottish Journal of Geology, 53, 41–52.

7 Fieldwork diaries 2018 We are taking over the world, Twitter, https://twitter.com/fieldworkdiary?lang=en Accessed 30 April 2018

8 Natural England (2018) Sites of Special Scientific Interest (SSSIs)

9 *Olivella mica,* World Register of Marine Species, Accessed 17 June 2018

10 Katharine E Welsh, Derek France, W Brian Whalley, Julian R Park Geotagging photographs in student fieldwork, Journal of Geography in Higher Education, 36, three, 469–480

14.7 References and resources

Anglia Ruskin University 2018 Guide to Harvard Style of Referencing, (6th Edn.), http://libweb.anglia.ac.uk/referencing/harvard.htm Accessed 15 January 2019.

Barley S 2012 What is a URL? YouTube, https://www.youtube.com/watch?v=SMxx9XEF6m0 Accessed 15 January 2019.

Cornell University 2018 How to Prepare an Annotated Bibliography, http://guides.library.cornell.edu/annotatedbibliography Accessed 15 January 2019.

Google Scholar 2018 Search Tips Google Scholar Help, https://scholar.google.co.uk/intl/en/scholar/help.html Accessed 15 January 2019.

Monash University Library 2018 Citing and referencing, Harvard (author-date) style examples, http://guides.lib.monash.edu/citing-referencing/harvard Accessed 15 January 2019.

Oxford University 2018 Managing your references: comparison tables, http://ox.libguides.com/c.php?g=423116&p=3851803 Accessed 15 January 2019.

UNSW 2018 Harvard Referencing for Electronic Sources, University of New South Wales, https://student.unsw.edu.au/harvard-referencing-electronic-sources Accessed 15 January 2019.

York University, Toronto 2018 Style guides for footnotes and bibliographies, http://researchguides.library.yorku.ca/c.php?g=679413&p=4789824 Accessed 15 January 2019.

Relaxed maths 2

We do maths all the time, without using a calculator solve:

350 ÷ 5 =	37 + 3 =	40 − 16 =
5 x 23 =	22 + 87 =	51 − 12 =
32 ÷ 4 =	11 x 16 =	103 − 71 =
74 + 16 =	21 x 9 =	201 ÷ 3 =

Answers in Chapter 28.

Answers: Try This 14.1 – Spot the referencing errors

1. This is missing italics, a comma should follow the brackets, and Publisher name is missing before the Place name. Also, Second Edition is written in full rather than 2nd Edn. Note: you won't lose marks using the full words, Third Edition, Fourth Edition consistently in your reference list. Consistency and accuracy is important.

2. This is the most common example of incorrect referencing found in student work: the order is wrong and there is missing data. In this example the author is also the publisher. It could start UN 2017 … or United Nations 2017 …

    United Nations 2017 *World Economic and Social Survey 2017*, United Nations, New York, https://www.un.org/development/desa/publications/graphic/wess2017-global-economy Accessed 15 January 2019.

3. Needs italics for the book title, 13th Edn. should be in brackets, and Location for Pearson is missing.

4. The Author's initials, Platform detail and the Accessed date are missing. Should read:

    Clapham ME 2015 Sediment Transport, YouTube, https://www.youtube.com/ watch?v=kKXd0dv6ae4 Accessed 15 January 2019.

5. The author details should not be in bold. *Area*, the title of the Journal must be in italics. The page numbers are included but the journal volume and issue number (50, 1) is missing. Should read:

    Korkiamäki R and Kallio KP 2018 Experiencing and practising inclusion through friendships, *Area*, 50, 1, 74–82, https://rgs-ibg.onlinelibrary.wiley. com/doi/full/10.1111/area.12352 Accessed 15 January 2019.

6. The authors' names are incorrect and include extra punctuation, add italics to Scottish Journal of Geology, and the URL or DOI:

    MacDonald JM, Magee C and Goodenough KM 2017 Dykes as physical buffers to metamorphic overprinting: an example from the Archaean– Palaeoproterozoic Lewisian Gneiss Complex of NW Scotland, *Scottish Journal of Geology*, 53, 41–52, http://sjg.lyellcollection.org/content/53/2/41 Accessed 15 January 2019.

7. This link will take you to the fieldwork diaries Twitter site, but not to the specific item with the name of the person who posted on site. More date information will help, but it might be difficult to find even with the date information because there are many posts.

    Fieldwork diaries 2018 We are taking over the world, Fieldwork Diaries @FieldworkDiary, Twitter, 5 March, https://twitter.com/fieldworkdiary?lang=en Accessed 15 January 2019.

8. Is missing the URL and Accessed date, and brackets are around the date. Underlining is not required. Should read:

    Natural England 2018 Sites of Special Scientific Interest (SSSIs), https:// designatedsites.naturalengland.org.uk/ Accessed 15 January 2019.

9. The Author and URL are missing. *Olivella mica* is correctly in italics as a species name but should not be underlined. It should read:

    WoRMS 2018 *Olivella mica*, World Register of Marine Species, http://www. marinespecies.org/aphia.php?p=taxdetails&id=448224 Accessed 15 January 2019.

10. This is a mess! The authors' full names are given rather than their family names and initials, the publication date is missing, and fonts are inconsistent:

Welsh KE, France D, Whalley WB and Park JR 2012 Geotagging photographs in student fieldwork, *Journal of Geography in Higher Education*, 36, 3, 469–480, https://www.tandfonline.com/doi/full/10.1080/03098265.2011.647307 Accessed 15 January 2019.

15

Ethics and plagiarism

The environment is everything that isn't me.
(Albert Einstein)

(BUT when and where did he say this? Is this plagiarism??)

Being ethical is the foundation of professional life. In all aspects of work there is an expectation of conduct that is based on moral standards which may be implicit or explicitly stated in a Code of Conduct. Most universities have a Student Code or Charter that provides the framework for ethical working practices. Matters of academic honesty are usually explicitly stated because there are penalties associated with violating the Code's expectation of honest academic behaviour. Behaving ethically is also referred to as having academic integrity.

Unethical activity, sometimes called academic fraud, is relatively rare but can carry very heavy penalties. Cheating in exams is clearly unethical and likely to lead to exclusion. Fraudulent activities include copying answers from other students without permission, paying other people to complete your work, lying about reasons for the non-completion of work or absence, smuggling notes into examinations, sabotaging or fabricating experimental results from laboratory and field work, and quoting other people's work without acknowledging the source (Chapter 14). Universities take all forms of academic fraud (unfair or unethical behaviour) seriously because it's vital that all students are assessed fairly and on the same terms as each other.

This chapter covers key points on plagiarism because students can struggle to understand this aspect of academic integrity. Plagiarising, presenting other people's work as your own, is unethical and likely to weaken the strength of your academic argument. Your essay is stronger for acknowledging the source of the ideas and case studies that you include. Let people know what you have read and utilised. Your aim is to be professional, ethical and honest as a researcher at all times. It may seem obvious, but making up or changing data to present a 'better' result is wrong. The other offences mentioned in the paragraph above are simply wrong. PLEASE also read your university information (try the website) on plagiarism.

15.1 What is plagiarism?

Umm, I think it's where you don't say who wrote something that you include in an essay

Here are some answers from universities in the UK and the US:

'... presenting someone else's work, in whole or in part, as your own. Work means any intellectual output, and typically includes text, data, images, sound or performance'. (University of Leeds 2019)
'Plagiarism is presenting or submitting someone else's work (words or ideas) intentionally or unintentionally as one's own, i.e. without acknowledgement'. (University of East Anglia 2018)
'When you demonstrate that you have done the research required to qualify you to join the conversation, you not only show respect for others' work, you also confer authority upon yourself and highlight the novelty of your particular contribution to the set of ideas under discussion. In these ways, citing sources represents a fundamental step in developing a scholarly voice.'
(Dartmouth College 2019)

This statement from Dartmouth College (2019) focuses on academic integrity and the nature of research as being part of the conversation amongst academics across institutions and generations. How would you summarise this direct quotation yourself?

Flinders University, Australia, helpfully identifies four behaviours to be avoided at all times:

'These behaviours will breach academic integrity:

- **Plagiarism** *is passing off someone else's work as your own. This includes copying word for word another's work, paraphrasing or using another's ideas as if they were one's own and "borrowing" facts and figures without proper acknowledgement.*
- **Collusion** *is when two or more students collaborate on an individual assignment. Examples of collusion include letting someone copy your answers on a test or allowing someone to write or edit your assignment.*
- **Cheating** *is copying answers on a test or commissioning others to write an assignment. Examples include using essay mills and ghost writers.*
- **Fraud** *is misrepresentation, such as asking someone else to sit your exam or falsifying data'. (Flinders University 2018)*

Ideas are all around us and students and researchers work with them all the time. The important thing is to acknowledge where ideas come from, and to place them in your words and framework. This is where the thinking part of searching and research (Chapters 8, 13 and 14) is important.

What is OK? – Summarising ✓, 'quotations with references' ✓, copying ✗

You are reading and citing research throughout your degree. The trick is to develop good habits when reading and making notes (Chapters 10 and 11) so that you do not plagiarise. Good practice involves reading a paper or chapter, and then **summarising** it yourself. This summary will usually be a shorter version of the original with all the important points included.

Read → Think → Summarise

Paraphrasing involves completely rewriting something in your words to convey the same message as the original. Just changing a couple of words in a paragraph is not paraphrasing.

Quotations are fine provided they are placed in 'quotation marks'. Quotations should be short and verbatim. Use them when the author's words are particularly important. To be clear that you are quoting an author exactly, inset the quotation from the normal margin. When you cite verbatim comments, from focus groups or interviews for example, please use italics. Note that student and staff quotes in this book are in italics.

Too many quotes is also a problem. An essay where 90 per cent of the words were correctly cited quotations in quotation marks, failed because the student had not demonstrated his understanding of the material, just used other people's words. In my experience quotations should be less than 5 per cent of an essay. Your own words show you have thought through and understood the material, demonstrating personal insights, which attracts marks.

Collaboration and collusion. Working with others, collaborating, is absolutely fine and an important part of group work (Chapter 16). If the assessment is a group report, sharing and writing together is important, and a great workplace skill to develop. BUT if asked for an independent report remember it is OK to work together on a project and share notes, but you must write your report in your words. Otherwise you may be accused of **collusion**, representing your colleagues' work as your own. Have a go at **Try This 15.1.**

Your university website may have a plagiarism tutorial or use **Try This 15.2.** Just do it, make sure you understand, and then you won't need to worry.

⚙ TRY THIS **15.1 – What is OK?**

Look at each of these examples in turn and decide what is acceptable academic practice. Then look at the answers at the end of the chapter.

1. Copying three sentences, changing their order, with no acknowledgement and no quotation marks.

2. Taking 200 words from a journal article, and putting it in the essay without any acknowledgement or quotation marks.

3. Paraphrasing the ideas from a paper in your own words, adding information from another source, and acknowledging both in the text, and the references at the end.

4. Copying three sentences from a textbook, changing two words, with no acknowledgement and no quotation marks.

5. Copying three paragraphs from a journal article, putting them in quotation marks, with the author's name and date, and the full citation at the end of the essay.

⇨ TOP TIPS

→ Every time you make notes ensure you include the full reference so you can put it in your final report or essay.

→ When adding a quotation in your notes, put quotation marks around it and add the full reference, so that you know to cite it properly in assignments.

→ Create a database of references (ask a librarian or tutor about software options).

→ Start working on a report or essay as soon as possible. Summarising well is only possible **IF** you have time to read and then think carefully and constructively. Allow your brain time to put ideas into a sensible order and come to some personal conclusions.

15.2 Websites, pictures, music – what can I use?

You cannot just assume that pictures, video, music and information on the web can be used in any way you wish. Many websites tell you what is copyrighted and what can be used, but the information may be elsewhere. Look at the front or home page of the website and the foot of the page. Your lecturers will cover Creative Commons licences in detail. Where a website states: 'all writing published on this website is licensed under a Creative Commons Attribution-Noncommercial-Share Alike 2.5 License', it means you can copy and distribute information from the site, as long as you acknowledge the source consistently and correctly every time.

This book cannot cover everything … Use **Try This 15.3** to get started.

⚙ TRY THIS **15.2 – Plagiarism activities online**

Many universities have exercises to show students exactly what plagiarism involves and how to avoid it. Take half an hour to check out these sites and your university's plagiarism resources. (All sites accessed 15 January 2019.)

Indiana University 2015 How to Recognize Plagiarism, https://www.indiana.edu/~istd/plagiarism_test.html

Learn Higher 2016 Preventing Plagiarism, http://learning.londonmet.ac.uk/TLTC/learnhigher/Plagiarism/index.html

University of Leeds 2019 Academic Integrity and Plagiarism, https://library.leeds.ac.uk/info/1401/academic_skills/46/academic_integrity_and_plagiarism

Colorado State University 2018 Plagiarism Self-Test, https://tilt.colostate.edu/integrity/resourcesStudents/quiz/

University of Technology Sydney 2017 Avoiding plagiarism tutorial and quiz, https://www.uts.edu.au/current-students/support/helps/self-help-resources/referencing-and-plagiarism/avoiding-plagiarism

⚙ TRY THIS **15.3 – Attributing copyright and authorship**

(all websites accessed January 2019)

Wikimedia Commons	Thousands of images, video and music that can be used freely with the referencing details attached to each image. See the Welcome page for details. https://commons.wikimedia.org/wiki/Commons:Welcome
Google Maps and Google Earth	Guidelines for non-commercial users: 'Generally speaking, as long as you're following our Terms of Service and you're attributing properly, we're cool with your using our maps and imagery; in fact, we love seeing all of the creative applications of Google Maps, Google Earth and Street View'. http://www.google.com/permissions/geoguidelines.html
Google images	Check out the 'Find Free to use images' pages. https://support.google.com/websearch/answer/29508
Wikipedia	Everything you need to know about copyright and use is at https://en.wikipedia.org/wiki/Wikipedia:FAQ/Copyright
NASA	On the media Usage Guidelines page see the information about non-commercial use. https://www.nasa.gov/multimedia/guidelines/index.html
Greenpeace	'Unless otherwise expressly stated, all writing published on this website is licensed under a Creative Commons Attribution-Noncommercial-Share Alike 2.5 License. This means Greenpeace UK keeps the copyright but you can copy and distribute articles from the site, as long as you acknowledge www.greenpeace.org.uk as the source'. http://www.greenpeace.org.uk/help/copyright
RSPB	http://www.rspb.org.uk/help/copyright.aspx
YouTube	https://support.google.com/youtube/answer/2797466?hl=en&ref_topic=2778546

15.3 How do lecturers spot plagiarism?

Academic staff mark hundreds, in some cases thousands, of student papers every year. They are very aware of writing styles as well as content. The easiest plagiarism to spot is when someone new to a particular topic is writing in an unsure manner, and then perfectly written, beautifully fluent paragraphs pop up. These are likely to have been written by somebody else. The most frequently missing references are for diagrams, maps and photographs. Their absence from essays, PowerPoint presentations and web pages is easily spotted by an examiner (Chapter 14.3 –14.5).

Tutors will conclude that your work is plagiarised when:

✗ it contains sections of text, diagrams or photographs without reference to their original source;

✗ a source is cited but there are no 'quotation marks' around direct (word for word) quotations;

✗ it has been bought from an internet website;

✗ it has been written by another student.

> *I really enjoyed reading the middle three pages of her report. It was elegant and well referenced. Then I remembered writing it myself, word for word, two years ago. Shame really …*

Plagiarism is spotted when people use unusual words. It is particularly easy to identify when students copy directly from their lecturers' books and papers. 'Always put direct quotations in quotation marks, followed by the author, date and page numbers' (Kneale 2019: 186).

Remember to be ethical in all your work. This is not just an essay and report problem. Plagiarism occurs in practical reports, maths and stats exercises, fieldwork reports and mapping exercises.

Many academic departments use software to identify plagiarism; Turnitin (2018) is one system. You will be introduced to this process, and usually have the opportunity to submit your writing for checking before submission. I very, very strongly recommend you go to the workshop that explains the process.

15.4 What happens when I am caught?

You won't be caught IF you never steal someone else's work. BUT for those daft enough to cheat, penalties are severe. Every website giving advice will emphasise that ignorance of the process is not an acceptable excuse. Every student is expected to understand the importance

> Passing off somebody else's work as your own is wrong. Expect to get caught and expelled.

of being honest in all their academic work. If in doubt ask your tutor, and include more rather than fewer sources. First year is a good time to make sure a tutor has properly explained the university expectations and procedures to your class.

Each university has regulations; see the university website and student handbook. When an incident occurs there is usually a local department investigation, followed by a university process. At each stage students have every opportunity to present their case. The students union is usually excellent at providing advice.

> Arguing that you did not understand what plagiarism is, is not an acceptable excuse.

A first offence from a first-year student may, possibly, be treated leniently if tutors are persuaded that you were unaware. Penalties typically range from repeating the work for the minimum pass mark, repeating the work to pass standard for a mark of zero, or being required to complete and pass the module but get a mark of zero overall. Whatever the penalty, it will affect your final degree marks, which does not look good on your graduation transcript. In second and third years, and for second offences penalties will range from loss of marks for the module to expulsion from the university.

Whatever the penalty your university imposes it is not worth getting into that position. Look up the message on plagiarism at Johns Hopkins University (2019): it has a 'zero tolerance policy'.

⇨ TOP TIPS

Avoid all the hassle involved in being suspected of plagiarism by:
➜ Starting assignments early
➜ Practising writing summaries
➜ Using quotations properly with '…' and adding complete references
➜ Running your work through detection software (Turnitin) before submission, if this is available

15.5 Ethical behaviour and data

There are courses in ethics, or parts of modules which address ethics and ethical behaviour in GEES degrees. It is especially important in dissertation work to ensure that any research undertaken that involves people or animals respects their rights. Obtaining ethical approval for a research project involving people is a normal part of the research process.

Collecting data, running experiments and fieldwork all take time. High-quality GEES research is produced by people thinking hard about their tests or experiments, and who are rigorous about the quality of the data they collect. Making up data is cheating. Universities take a dim view of students who cheat. Students who are caught have usually been short of time to do the work properly. This is why dissertations are usually six- or nine-month activities, allowing time to design and collect data, analyse the information, think about the results and decide what they mean. Consider Figure 15.1. Why might these graphs ring alarm bells for a marker?

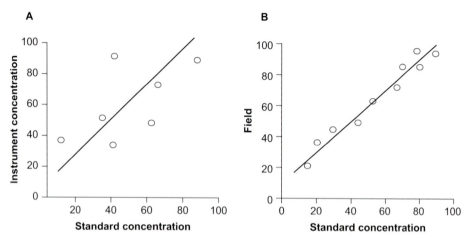

Figure 15.1 (A) Laboratory calibration of an instrument against chemistry standard solutions; (B) Range of results recorded in the field experiments.

Graph (A) shows the laboratory calibration for an instrument against a standard, graph (B) shows the field results. Where an instrument shows variation when tested with laboratory standards it is bound to be recording with greater variability in the field. People who cheat tend to present data that is too perfect, too good to be true.

It can be really hard to get people to fill in questionnaires, especially when standing in a town centre in the rain. The temptation to get 15 answers and make up the next 15 is high. Please don't do it. Tutors will ask to see the raw data, look at the handwriting, and will notice if amazingly you did five questionnaires in five hours on three days, and then completed 40 questionnaires on day four. Tutors realise how difficult it is to get good-quality data and answers. Explain why your data is limited in your report. Be cautious about the interpretations you make from it. Cheating by falsifying data just isn't worth considering.

It is normal in any dataset to find anomalous results. These come from human error in collecting the data, errors in instrumentation, errors in entering the data for analysis, and … Most field values are highly variable, with hourly, daily and seasonal variations, as well as point-to-point variability. GEES data is always going to be messy. The fun is in interpreting the muddle. Generating completely clear conclusions is likely to ring warning bells for markers. When there are some anomalous data points, comment on them – please don't delete them. It is proper to suggest you might have made an error in data collection or data entry. It may be there are natural data variations of which you are unaware but which your tutor expects. For example: a student replaced the recorded pH data at station x because it was out of line with all the other values, but the geology changed at that point. The tutor expected to see very different pH levels at that station and identified the cheating.

Always keep all your data sheets, questionnaires, and notebooks until after the research is finished, reported and marked so that you can check back to your original data.

In some departments there are student research projects where data are shared across years; your dataset includes your results and results from students in previous years. Then your results become part of this larger dataset for next year's students. Keeping accurate records is obviously very important in this case, but equally important for all projects. Being honest in reporting results shows sound judgement as a scientist or social scientist. You want this reputation.

15.6 References and resources

Burkill S and Abbey C 2004 Avoiding Plagiarism, *Journal of Geography in Higher Education*, 28, 3, 439–446.

Cambridge University 2019 Why does plagiarism matter? https://www.plagiarism. admin.cam.ac.uk/what-plagiarism/why-does-plagiarism-matter Accessed 16 January 2019.

Dartmouth College 2019 Sources and Citation at Dartmouth College, https://_ writing-speech.dartmouth.edu/learning/materials/sources-and-citations-dartmouth Accessed 16 January 2019.

Flinders University 2018 Policy on Academic Integrity, http://www.flinders.edu. au/academicintegrity/student.cfm Accessed 16 January 2019.

Johns Hopkins University 2019 Notice on Plagiarism, http://advanced.jhu.edu/ current-students/policies/notice-on-plagiarism-2/ Accessed 16 January 2019.

Turnitin UK 2018 QuickStart Guide, Student Guides, https://guides.turnitin. com/01_Manuals_and_Guides/Student_Guides/01_QuickStart_Guide Accessed 16 January 2019.

University of East Anglia 2018 University Policy on Plagiarism and Collusion, https://portal.uea.ac.uk/documents/6207125/7465906/Section+3+Plagiarism+ and+Collusion.pdf Accessed 16 January 2019.

University of Leeds 2019 Academic integrity and plagiarism, https://library. leeds.ac.uk/info/1401/academic_skills/46/academic_integrity_and_plagiarism/ Accessed 16 January 2019.

Geo-paths 3

Can you find ten or more words in the square? Move horizontally and vertically, but not diagonally, from letter to touching letter. Each letter can only be used once in a single word. There is one word that uses all the letters. Answers in Chapter 28.

E	R	E	P
P	T	N	I
R	E	U	H
E	N	R	S

Answers: Try This 15.1

Points 1, 2 and 4 are absolutely unacceptable plagiarism. Point 3 is OK. Point 5 is lazy work. It may not be deemed to be plagiarism, but expect a low mark for not thinking through and developing your interpretation of the information. Where you have embedded a long quotation, the best advice is to paraphrase and re-present it.

16

Collaborative learning: team research and reporting

How do you know what you think before you talk about it?

Group work is a fundamental part of school and university education. Employers want people who work happily and effectively in teams to solve problems, issues and challenges that need the co-operation of many people. Teamwork allows the exploration of more material than is possible for an individual, which at university means you can look at more interesting and complex issues. Team members bounce ideas off each other. Where resources are limited, getting involved in co-operative activities and sharing has benefits for everyone. The skill benefits of teamwork include developing professional standards in presentation, group reporting, problem solving, negotiation, confidence and responsibility.

A group of students having supper together may talk about Prof. Blownitt's lecture on nuclear power. Everyone has different ideas. If some people decide to chase up further information and pool it, they are beginning to operate as a team. In some modules you will research and write as a team, but remember you can use team skills and approaches to tackle other parts of your course. Two to five people in the same flat, hall of residence or tutor group can team up to extend and optimise their research activities, and talking about geography or geology together will develop your discussion skills and broaden your views.

16.1 Teamwork tactics

This chapter raises some of the issues associated with group work and suggests some ways of tackling group tasks. There are pitfalls. Not everyone likes teamwork and sharing. Some people feel they may be led down

> *Where is …? We said meet at 10 after the lecture. Text him!*

blind alleys by their team-mates. Overall, I think the advantages of team research outweigh the disadvantages, but if this bothers you, brainstorm reasons why certain group members are a disadvantage. Possibly:

☹ those who do not pull their weight;
☹ the perennially absent;
☹ those who do not deliver on time;

☹ the overcritical who put others down, suppressing the flow of ideas;

☹ anyone who gets touchy and sulks if their ideas are ignored.

To benefit from team activities these are characteristics to suppress! You may not like working in teams, but when you do it is in everybody's interests for the team to pull together and derive the benefits. Having identified unhelpful teamwork characteristics, consider what good team qualities might include:

☺ making people laugh;

☺ focused on getting stuff finished;

☺ sharing information;

☺ proposing suggestions;

☺ communicating well;

☺ being reliable;

☺ making friends;

☺ keeping calm in arguments;

☺ being logical and systematic;

☺ speaking your mind;

☺ resolving disputes.

> Positive phrases encourage you and the group:
>
> *Thanks, really interesting.*
>
> *I need to think about that.*
>
> *Really helpful.*
>
> *Wow/Cool/Wicked.*
>
> *Great idea.*
>
> *That's a really useful insight.*
>
> *Brilliant, we can use that.*

What else? Reducing tension in a group promotes cohesion and encourages everyone to learn. Taking this positive approach is known as appreciative enquiry. Use Figure 16.1 to plot how conversations, plans and ideas evolve. Ask: What is in progress? What have we achieved? What is successful? What else? Tracking helps to identify and encourage using positive approaches which can help everyone succeed.

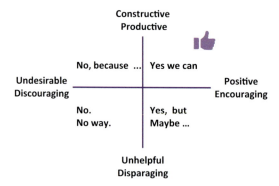

Figure 16.1 Tracking the characteristics of thinking and discussions

There is a natural pattern in life that can also be seen in the normal reaction of people to teamwork and life in general (Figure 16.2). Map these natural reactions to an event in your life, or to an assignment, see example in Figure 16.2. Each time you are given an assignment, you are likely to experience all these reactions.

The people who realise 'where' they are in the sequence, and push on to the positive recovery and getting-on-track stages, give themselves more time to do the task well.

Groups can explore and evaluate more material than an individual, doing more interesting projects; but sharing research ideas and research outcomes can be difficult. People work at different speeds, and have different demands on their time. Group members are only human. Everyone has a different way of researching and note-making; shared information 'looks' different on paper, BUT you can learn from the way other people note and present information. Figure 16.3 summarises thoughts from three groups after a shared research task. They give an insight into what can happen.

> My pre-uni group work was not a good experience. I don't like it. I know where I am if I do something myself, but my module group was great and it was so much fun, getting to know the others … and I now understand about t-tests because Emily explained them to me. Looking forward to next time.

Natural reaction	Reactions mapped to an essay	Reactions mapped to a group project
Shock	I know nothing about … *OK I have done projects before …*	We know nothing about …*Help – who knows about …*
Recrimination	I don't want to do this module/essay, why is it compulsory?	I don't want to work with: this group/people/at all!
Disagreement	No motivation, not much done. *I have done some bits and pieces, it's good to have started!*	Everyone is doing their own thing or nothing. *Lots going on, (I think), we need to share …*
Reorganisation	Two weeks to go, time for a plan. *I like doing plans, and that will get me going …*	Two weeks to go, we must make a plan for and with everyone. *Sharing what we have already and dividing up the jobs is fun. I am avoiding … and doing the bits I like.*
Recovery	Parts of this are quite interesting.	Everyone knows what to do, some people are enjoying it.
Getting on track	Parts of two sections are organised, and am finding more references.	It's coming together, drafts are being swapped around, and ideas developed.
Partial success	Second draft done, needs editing, good diagrams and references. *Great, nearly there, ticking off the last jobs …*	Mostly done, but no cover, diagrams missing, references incomplete. *Useful meeting, shared out the finishing off jobs …*
Frustration	Printer needs ink, spell and grammar check to do.	We cannot find Jim who has it all on his USB stick.
Success	Final version handed in. Hooray! *Party Time.*	Report completed, time to celebrate. *Party Time.*

Figure 16.2 Natural reactions to group work. Moving from less to *more helpful* thinking. Some *positive thoughts in italics* – add more positive thoughts to all sections.

Task	Process
Tutorial essay on wind power Individual essays Two-week deadline	*Met and decided we didn't know where to get data! Split up to do library, online library and www search. Discovered power company brochures and reports. Agreed to keep textbooks and photocopies in Ed's kitchen cupboard where we could all get at them. Middle of week 2 we realised no one had looked at Wind Power alternatives, Sarah did it. Two discussions at lunchtime in the Union, and it became the topic of conversation in Ed's kitchen for a week. Sarah and Mike phoned two companies for further details and got some info, but too late for the company to post things to us before the deadline.* *The research time was about equally split. Generated a bigger pile of articles at Ed's than we could have found alone. Results were a bit general. We needed a better plan at the start or part way through week 1. Found info in the Engineering Library at the end which we should have used, it would have helped. We got lots of electronic source info, which saved time, BUT it cost a lot of printer credits. In future we should be more selective and email to each other to save printing.* *Good to chat, group got to know each other better. One person thought she would have started writing sooner, but waited for info from others.*
Seminar preparation Four people Group presentation for 20 minutes followed by general discussion, class of 25	*We all knew each other and it was easy to share out the reading, we volunteered to cover different bits. We decided to meet 5 days before the seminar, over supper, to discuss who was doing what. Great evening, great discussion, no one wrote anything down, so we had to meet the next day to draft the talk.* *We had far too much material but we could choose. Took examples from Bosnia (it's topical) and Sudan (map and pictures available). Put other refs. on a handout. Decided there were some good www sites and added site details to the handout.* *Decided to use an 'awkward questions' format. We took it in turns to ask an awkward question, and had two people who tried to answer each one. Changed the questions a bit! Met the day before to practise. Everyone talked for too long, tried to cut bits out. The introduction changed about twenty-five times!* *The seminar went OK, mostly because we had talked about the points beforehand, so I think we all knew so much more than usual.*
Organising a field research day Five people Task to be decided	*Sorting out what and how to do it took ages, there were too many good ideas. We had trouble getting together, and in finalising the plan. Tended to split into two. The organisation of the day was left to Andrea in the end, because she had the car. We met three times beforehand and didn't really decide anything each time. We needed a leader from the start and to focus faster on what to do.* *The field day was great FUN, Liz and Jamie put the data straight onto the laptop. We changed what we were measuring a few times so by the end we had more records for the sites visited later in the day. Didn't get as much data as we hoped, but talked to some really nice people and the graphs look good. We had loads to write under the 'criticism' and 'what we would do next time' sections.*

Figure 16.3 Experiences of group research

One meeting or three? How often does a group need to meet? Generally: two or more meetings ➔ more chat ➔ greater exchange of views ➔ more interactions ➔ more learning. Have electronic meetings when time is tight. Facebook, VLE chat rooms, Skype and email all work well with some practice.

16.2 Writing in teams

Team writing is a normal business activity using the same techniques as for projects and reports. Start with the team brainstorming an outline for the document and deciding who does what; then add individuals' research and draft subsections; circulate

For group messaging:
→ Be polite and friendly.
→ Remember everyone will have different views, opinions and information. Double-check you have attached the right (latest) version of the document.
→ The group writing process is a muddle at times, just as your own research notes are a muddle before you organise an essay structure. Create time to organise and structure the work.
→ Reply to messages within the day, even if it is just to explain you will do the next bit tomorrow. Keeping in touch is important for good group dynamics.

subsection drafts for comment; incorporate additional ideas and views; and finally someone edits the final version. Like ensemble recorder playing, writing as a team is difficult and discordant the first time you do it, but a very useful experience.

One valuable approach involves team members volunteering both to research and draft specific subsections and taking agreed 'writing roles' (Figure 16.4). Discussions are livelier and more focused when individuals can use their role to offer alternatives and new approaches. The 'critic' has the license to criticise, the 'linker' can say 'yes that's great, but how do we use butter mountains and polar ecology to explain county employment patterns?' Adopting writing roles depersonalises criticism, which is especially important when working with study buddies. Be critical and stay friends.

Writing role	Responsibilities
Graphics and pictures	Smart graphs, figures, pictures, indexes, contents page
Summari ser	Introduction, conclusions, abstract
Critic	Faultfinder, plays devil's advocate
Academic content	Subject reporters
Linker	Checks connections between arguments, sections and the introduction and conclusion
Discussant	Evaluates and discusses
References	Checks the selection is complete and in the agreed consistent format

Figure 16.4 Writing roles

Style and layout

Decide on the general layout of the document at the start, the format, fonts and style of headings, and the format for references. Decide on word length, for example 'each subsection has a maximum of 200 words'. Early word-length decisions limit waffle. Nevertheless, length needs looking at later in the project, as some parts of the argument will need more space. Keep talking to each other, share ideas and discuss their relative importance in the narrative.

> *It's all feedback. I think that's what I didn't understand when I came to uni. Comments from tutors are feedback, but so are comments and ideas from the rest of the group. Cool.*

Finally, someone (or two) has to take the whole document and edit it to give it a consistent voice and style, BUT everyone must provide them with the graphs, diagrams and references in the agreed formats.

Duet writing

Two people sitting together can be very effective. The exchange of ideas is immediate, two brains keep enthusiasm levels high and you plan further research activities as you go. It is particularly advantageous to have two people doing the final edits together, keeping track of formats, updates and staying cheerful.

Timetabling issues

A team-writing activity cannot be tackled like the traditional essay, on the night before submission after the pub. People get ill and things happen, so the timetable needs to be generous to allow for slippage AND team members have to agree to stick to it. It usually works, as it is too embarrassing to be the only non-contributor. Put completion dates against decisions, as in Figure 16.5, and use Post-it notes to create a picture version of this list. (You can move Post-it notes to different days when delays happen and ideas change). Which approach suits the group better?

Keeping a team on track? Need a chairperson?

One of the pitfalls of working with friends is that more time is taken making sure everyone stays friends than getting on with the task. One vital issue is the initial division of labour (see Figures 16.3 and 16.4). Everyone should feel happy, involved and equally

1. Brainstorm initial ideas, assign research tasks and data collection (Day 1)
2. Research topics and collect data (Days 2–5)
3. Individuals draft their subsections and share them around the group (Days 6–9)
4. Meet to discuss progress, decide on areas that are complete, assess where additional research is needed, assign further research and writing roles and tasks (Day 10)
5. Individuals redraft sections and circulate them again (Days 10–14)
6. Decide whether we are all happy with this? Pass final version to the editor (Day 15)
7. Team members do final checks on figures and references (Days 16–18)
8. Finalise cover, contents and abstract. Everyone has a final read. Check and add page numbers. Check submission requirements are met. (Day 18)
9. Submit (Day 18!)

Figure 16.5 Timetabling team writing

valued. So the chairman must endeavour to en-sure there is fair play, that no one hogs the action excluding others, and equally that no one is left out (even if that is what they want). The chairman is allowed to hassle you into action. It's unfair to dump the role of chairman on the same person each time – share it around. Chairing is a skill everyone should acquire.

> *We did seven group projects this term. All really good fun when we had sorted out what to do. I had to have a diary for the first time. Lots of meetings. What I learned is that something always goes wrong, NIGHTMARE! it's just like that. But we got them all done for the deadline.*

If a group feels someone is a serious dosser, they may want to invoke the 'football rules', or ask the module tutor to do so. The rules are one yellow card as a warning – two or three yellow cards equal a red card and exclusion from the group. The yellow–red card system might be used to reduce marks. A red-carded person does not necessarily get a zero, but in attempting to complete a group task alone, they are unlikely to do better than a bare pass.

⇨ TOP TIPS

✓ The time required to tidy up, write an abstract, make a smart cover, do an index, acknowledgement page and cross-check the references is five times longer than you think – split the jobs between the team.

✓ It saves hours of work if everyone agrees at the start on a common format for references, and everyone takes responsibility for citing the items they quote.

✓ The key to team writing is getting the STYLE and TIMETABLE right. (CONTENT also matters!)

Unethical behaviour

Where group work leads to a common report then obviously collaborative writing is involved. When writing independent reports ensure each is written independently (not copies or cut-and-paste versions of each other's documents). In this situation, share reading and discuss ideas, BUT write independently. Plan to finish the research five days before the deadline so everyone has time to draft and revise their own reports. (Chapter 15).

16.3 Communication by email, team sites, SharePoint, Google docs, Skype …

It is vital that everyone has access to a group report, so agreeing the process for sharing and updating the most recent version of a document is crucial. Using email and team sites with attachments is ideal and saves printing. Your university will have various options, and Skype and similar platforms are great for discussions when you cannot meet together. Sharing online lets you work at convenient places and times.

Imagine a report constructed by four people, being edited and updated daily. Team-writing software will take care of some elements of the list below. But until then a tracking system is crucial. Adapt elements of the following to suit your project.

The document needs:

✓ A header page with title and the outline plan.

✓ An agreed working order (and timetable), e.g. John drafts Section 2, then Anne revises it. All revisions circulated by day X.

✓ A section that everyone updates when they change the document, e.g. 'Andy modified Sections 2.6 and 3.2 on Wed 1 April at 10.00'.

✓ An agreed format for citations and reference list.

✓ Agreement to check email and respond every X days.

✓ Decision to use the word-processing package 'Revision Editor'. The person responsible for section X will not necessarily welcome five independent redrafts. The 'Revision Editor' highlights revision suggestions for the subsection editor to accept or reject as desired. By opening multiple copies of the document on your pc, you can cut and paste between drafts.

✓ Use section numbers and date each version carefully. Page numbers change with almost every version, too confusing.

✓ Agree that everyone will save all drafts, so that you can revive an earlier version if disaster strikes, people are ill, lose their USB stick …

Use email (Skype, Facebook …) to brainstorm ideas among friends, tutorial and seminar groups and old school friends doing degrees at other universities. People like getting messages and usually respond. There are a great many people you could brainstorm with; think more widely than just you and your lecturer.

16.4 Assessing team reports

Assessment tends to generate 'fairness' conversations. There are dark hints about cheerful dossers getting good marks when their mates have done the work. How is this handled? Most staff will offer some variation on the following approaches for assigning marks, some of which involve team input. If the 'football factor' is in play, there may be an agreed penalty, for example −10 per cent for a yellow card. There are different approaches to assigning marks:

The simple approach – everyone gets the same mark, so it is up to everyone to play fair.

The private bid – each individual completes a form privately, and the assessor resolves discrepancies:

This approach lets people with personal problems acquaint staff privately. It can require the Wisdom of Solomon to resolve. Remember that lecturers always take careful note of what you say, but do not necessarily change the marks.

Name............................. ....
Names of other team members   ....
I feel my contribution to this project is worth...........% of the team mark
because.. ....
....
Signed.. Date..

Team input – everyone comments on the contribution of each team member. Summing the totals constructs an index of activity, which is used to proportion the marks:

Estimate the effort made by each team member, 0 = no effort, 2 = did a bit, 3 = average, 4 = really useful, 5 = outstanding contribution.	Alex	Sara	Jazz	Mark
Attended all meetings				
Contributed ideas				
Did agreed share of the research				
Did agreed share of the writing				
Other information – please detail particular contributions:				

Students completing team assessment contribution forms are usually ethical and honest when describing complications, admitting if they have caused problems and suggesting where marks should be raised for people who did extra work. It is the right thing to do. You will get raised marks when you have helped in a crisis.

16.5 Student and lecturer feedback

How do these thoughts (feedback) from a lecturer match your experiences?

'Initially students were very democratic and gave equal marks. Later in the year, with more experience, students raised the marks of those who had done more of the task'.

'The amount of chat was enormous, and between them they found they could tackle a
  more difficult problem than as individuals'.

'The team-writing exercise made everyone think about the order and quality of the
  material. The report was more extensive and detailed than most work I see, because it
  had been drafted and redrafted, more thoughtful work'.

'One team was happy to include someone who is generally absent and not bothered
  about doing well because, as they explained, s/he happily did the non-academic stuff
  like washing up in the lab, getting drinks and making poster backgrounds while
  cracking jokes and keeping the team cheerful'.

Now look at the student feedback in **Try This 16.1**.

TRY THIS 16.1 – Student reflections on group research

What are your thoughts? Are there tips and tactics here that you can use in future?

What were the advantages of teamwork?	What were the disadvantages of teamwork?
People can help each other to understand. Learn consideration. Lots of prior knowledge got us started quickly. New friends/Met new people. Less work per person. Shared ideas. Divided workload saved time. Created cohesion amongst five people who had only just met. Greater overview of the subject. We covered library, web and e-journals between us, more than we would have managed alone. Learning to share and compromise. Individuals could concentrate on the bits that interested them. Encouraging to know that others in the team thought you were on the right lines.	Recognising that other people have different ways of working, and having to find a way round it. Hard to find out if a non-contributor was ill or skiving (was ill). Who does the writing? Difficult to agree on points of view. Hard to get a time to meet f2f, OK-ish online. Conflicting views (is this an advantage?). Having to compromise on common topics.

What will you do differently next time you work together?

More thorough planning to concentrate efforts. Set clear objectives at the start.
Improve online sharing to save time. Email each other more often. Read my emails regularly.
Elect a team co-ordinator, who has the team members' timetables, to find convenient meeting times.
Split the team to tackle different libraries and meet later.
Organise a couple of meetings rather than trying to do it all in one go.
Reserve books in advance, arrange to meet in library at a quieter time of day. Search online first, papers and books later.
Co-ordinate reading, so two people do not read the same text.
Start work on the project earlier – not leave doing this to the last three days.
Distribute the workload earlier.

16.6 Umm, I can't be there … help!

Over the course of the degree, with 30 or 40 group activities something will go wrong on one or two of them. That is true for everybody. When there is an emergency people are good at reorganising and coping; it's a project management skill.

Group work involves sharing tasks and managing problems as they arise. Tutors are deeply unimpressed by people who do not give their team fair warning of problems, so if you have a major commitment during group work explain at the planning stage. Arrange to do your share when you can,

> As a group we worked really well. Sal had to go to her sister's wedding the last weekend before Monday hand-in. She did loads of the research, sorted diagrams and left us with probably more stuff than we would have found. Sweet. The rest of the group did the final editing and checking.

more of the early or finishing tasks. In an emergency, you are ill or … email or text the group. Tell your tutor so s/he can support you and the group.

16.7 Evidencing team skills in job applications

Make the most of group-work activities on your CV. Ignore the academic content, focus on the skills you used: negotiation, meeting deadlines, allocating tasks, collaborative writing, editing and co-ordinating research. You've done the work and got the experience, use it effectively in applications and interviews. Adapt the following sentences to record your involvement:

☺ *Project … gave me the opportunity to practise giving feedback as well as hearing what the team thought about my contribution. As a result we/I have …*

☺ *Tackling/achieving … was possible because the team had the benefit of the experience of … and contributions from …*

☺ *What I learnt from … was that everyone's opinion is different and you have to agree on a particular course of action. To do this we/I …*

☺ *What I've learned during the course of the last few projects is that we won't have a good outcome unless we get behind the chosen route, even though not everybody agrees it's the best way …*

☺ *The best thing about teamwork is when someone in the group has a completely different idea, something I've never thought of, and we get behind it, for example when …*

☺ *… went horribly wrong because we didn't get the end dates sorted for each part of the project. And then no one really knew what was happening. I've taken on monitoring dates in recent projects and really enjoy that part of the process. Keeping track by … allowed …*

The main points to note after a project are:
✓ What went well and why.
✓ What didn't go quite to plan and what we did instead.
✓ What had to be adjusted and how we did it.
✓ Learning outcomes: what to do differently next time at the three main stages: planning, implementation and writing.

16.8 Resources

Hay I 1999 Writing Research Reports in Geography and the Environmental Sciences, *Journal of Geography in Higher Education*, Directions, 23, 1, 125–135.

Dawson M 2013 Group Project Management, http://www.learnhigher.ac.uk/working-with-others/group-work-working-with-others/group-project-management/ Accessed 16 January 2019.

Open University 2011 Groups and teamwork, OpenLearn, https://www.open.edu/openlearn/science-maths-technology/engineering-and-technology/groups-and-teamwork/content-section-0 Accessed 16 January 2019.

University of Birmingham 2018 Why work in Groups? https://www.birmingham.ac.uk/schools/metallurgy-materials/about/cases/group-work/why.aspx Accessed 16 January 2019.

University of Kent 2018 Careers and Employability Service Team working skills, https://www.kent.ac.uk/careers/sk/teamwork.htm Accessed January 2019.

Words in geo-words 2

Set the stopwatch for five minutes on your phone, tablet or computer. How many words of three letters or more can you find in

PALEONTOLOGY?

☺ Thirty is a good score. Answers in Chapter 28.

Discussion

Why is it that when you discuss Murphy's Law something always goes wrong?

Academics spend much of their time working collaboratively, discussing their research ideas because testing information and development as a researcher primarily happens during these discussions. Talking through the details of a topic leads to a greater degree of understanding and learning as ideas are shared, tested, revised and refined. Consequently modules have lots of discussion opportunities (workshops, tutorials and seminars) where ideas are created and innovated. Discussions vary from very short, for example in a lecture where you discuss something with a neighbour for a couple of minutes, to full-scale seminars with a 5-minute scene-setting presentation that starts a 40-minute discussion. Tutors will expect everyone to have read specific materials before a session (sometimes called flipped classroom) so that everyone is comfortable understanding at least some of the material from the start. Being a confident discussant pays off later because most jobs involve discussing topics calmly, fairly and persuasively (*professionally*).

To learn effectively from discussions a relaxed atmosphere is vital. Getting to know people is important, so some social chat at the start is always good. Ideally work in a thoughtful calm atmosphere, with silences that give people time to think about content and make notes, but excited works too. Being comfortable with silence while people think is a very valuable skill.

Making notes during discussions (pad, phone, tablet) is crucial: you capture ideas as they whiz around and sort out what you want to say. Ineffective discussions occur when people are worrying too much to listen; mentally rehearsing their lines rather than concentrating on the speaker. Some of this is nerves which will calm down with practice, but in the meantime preparing fully is the best way to lower stress levels (*do the pre-reading*). You have background and specific information to share.

Aim to actively seek the views of others and value their contributions. Employ open-ended questions, those which encourage an elaborated, rather than a brief yes or no answer. 'What are the main resort facilities of Penang?' is an open-ended question, more useful than 'Do you like Penang?' 'What do you think about green issues/global warming/transport in New Delhi?' are good questions, but 'What do you think ...' is a bit general. More useful phrases include:

'What are the advantages of ... [global warming]?'
'Do you consider ... [the internet's global, unrestricted spread influences the actions and attitudes of individuals]?'
'What should be valued about ... [the lake and its environs as a community recreation resource]?'
'What is the evidence for ... [linear lava fountains constructing pyroclastic cones]?'
'What are the causes of ... [peatland destruction in the Amazon]?'

Keep up the quality of argument in discussion (*include references*). For example, discussing river sediment management, you might make a general point: 'There are natural variations in sediment loads and fluxes

> What you are really saying is you want us to discuss like we would write in an essay?

with time and changing environmental conditions'. You could be more specific: 'plans in place to manage sediment flows following the eruption of Mount St Helens 40 years ago, haven't worked and are having to be replaced'. This is more specific and touches on an example, but the discussion point (argument) would be strengthened (more marks) by adding references and more detailed examples as you speak. You might say: *'Sclafani, Nygaard and Thorne in their 2017 paper state that the flow of sediment in the rivers draining Mount St Helens is unpredictable, so that the management plans need to be kept under review and regularly revised. In their conclusions they emphasise that the engineering and geomorphology principles and practices can be transferred to other catchments, but managers must recognise the uncertainty and natural variability in these systems that makes forecasting potentially ineffective'.*

> YES, use technical terms and references. That's what makes a strong academic argument, oral or written.

This has added more technical information, and demonstrated your research skills. Take a look at the paper and consider how you would paraphrase their narrative for discussion.

17.1 Types of discussion

Brainstorming

Brainstorming is a great way of collecting a range of ideas and opinions and getting a group talking. The process involves everyone calling out points and ideas. Someone keeps a list on a whiteboard or flip chart so everyone can see. A typical list has no organisation; there is overlap, repetition and a mix of facts and opinions. The art of brainstorming is to assemble ideas, including the wild and wacky, so that many avenues are explored. The points are reordered and arguments developed through discussion, so that by the end of a session there is a pooled, ordered and critically discussed list. Use **Try This 17.1** as an example of working with a brainstormed list, and **Try This 17.2** for the next essay.

> *It seemed very ... uncool to make notes about what your mates say. ... swotty. I should have started doing it sooner. I missed loads of good stuff at the start ..*

TRY THIS 17.1 – Working with a brainstormed list

Brainstorming produces a list of ideas with minimal detail and no evaluation. This list was compiled in five minutes during a coastal field trip. Items overlap and repeat, facts and opinions are mixed, and there is no order. Take five minutes to categorise the items. An example sort is at the end of the chapter.

Physical and human influences on the coastal landscape at..............:

Concreting the slopes prevents soil erosion.	Repairs to sea wall.
Rocks at cliff foot dissipate wave energy.	Tourist facilities destroy foreshore view.
Tourist facilities destroy foreshore view.	Tennis courts on old levelled slip.
Tree planting on recent and older slip hitting sites.	Undercutting of cliffs by waves.
Vegetation worn on paths and alongside paths.	Movement of handrails.
Caravan park on cliff.	Path covered by slumping clay.
Bolts into the rock face increase stability.	Car parking unscenic.
Soil overlying clay.	Groynes prevent longshore drift.
Sea wall prevents undercutting.	Car pollution.

Fine and coarse netting, geotextiles, used on clay slope to help stabilise the soil.
Rock armour installed at the foot of recent landslip.
Complex geology, land slips, big fault structures.
Evidence that paths across the slips are regularly re-laid, cracked tarmac, fences reset.
Drainage holes in retaining walls to reduce soil water pressure.
Car parking demands, unsightly in long views.
Litter needs collecting from beach. Road built over old stream, presumed piped underground.
Castle fortification, some walls now undermined and eroded.
Hotel, amusement arcade and tourist shops – noisy and unsightly.
Varied quality of coastal path, not always well signposted.
Café for tourists and visitors at foot of slip area.
Harbour has mix of marina, fishing and tourist fairground rides.
Victorian swimming pool, in need of repair.
Wall to prevent boulders from cliff hitting the road.
Promenade held up by an assortment of stilts, walls and cantilever structures where slope below has been undercut.

Using your next writing task as the focus, brainstorm a list of ideas with colleagues and friends, including references and authors, and use this discussion as the basis for essay planning. Brainstorming 'what I know already' at the start of planning indicates where further research is required.

Role-play exercises

Role-play is used in modules to help you to understand the multiple dimensions of many issues, the different parts of complex jobs and processes. And they are so much more fun than reading. Role-plays can involve: running a community consultation meeting, running a local government consultation meeting, bidding for funds for …, presenting a project proposal to a board, presenting options for decisions by a planning authority, a press conference …

The planning game was great fun, cos we all had different roles and different points to make. It made the arguments much more real, some people got really excited, especially the protesters group. It made me think of how people who live there would look at our designs.

   A good role-play will involve preparing the academic arguments for or against (a development in a national park, drilling a borehole, starting a business) and presenting them in a role (consultant, engineer, lobbyist, pressure group representative, councillor, MP), not necessarily a role you would agree with personally.

Debate

The normal format for a debate involves a contentious issue presented as a motion. It is traditionally put in the form: 'This house believes that …' One side proposes the motion and the other side opposes it. The proposer gives a speech in favour, followed by the rebuttal presenter speaking against the motion. These speeches are 'seconded' by two further speeches for and against, although for reasons of time these may be dispensed with in one-hour debates. The motion is then thrown open so everyone can contribute. The proposer and the rebuttal presenter make closing speeches, in which they can answer points made during the debate, followed by a vote. Since GEES issues are rarely clear-cut, a formal debate is a useful way of exploring positions and opinions, and for eliciting reasoned responses (put *references* in your speech).

Oppositional discussion

Oppositional discussion is a less formal version of debate, in which each side tries to persuade an audience that a particular case is right and the other is wrong. You may work in a small group, assembling information from one point of view, and then argue your case with another group which has tackled the same topic from another angle. Remember that your argument needs supporting with evidence, so keep case examples (*references*) handy.

Consensual discussion

Consensual discussion involves a group of people with a common purpose, pooling their resources to reach an agreement. Demonstrations of good, co-operative, discussion skills are rare; most of the models of discussion on TV, radio, and in the press are set up as oppositional rather than consensual. Generally you achieve more through discussing topics in a co-operative spirit and, one of the abilities most sought after by employers of graduates is the ability to solve problems through teamwork.

Negotiation

Negotiation, coming to an agreement by mutual consent, is a really useful business skill. People practise and improve their negotiation skills every day – by persuading a tutor to extend an essay deadline, getting a landlord to repair the heating or someone else to clean the kitchen. In formal negotiations aim to:

✓ Prepare by considering the issues in their widest context.
✓ Enumerate the strengths and weaknesses of your position. It reduces the chances of being caught out!
✓ Get all the options and alternatives outlined at the start. There are different routes to any solution and everyone needs to understand the choices available.
✓ Check that everyone agrees that no major issue is being overlooked, and that all the information is available to everyone.
✓ Appreciate that there will be more than one point of view, let everyone have their say.
✓ Stick to the issues that are raised and avoid personality-based discussion; s/he may be an idiot, BUT saying so will not promote agreement.
✓ Assuming decision deadlines are flexible, break for coffee or agree to meet later if discussion gets overheated.
✓ At the end, ensure everyone understands what has been decided by circulating a summary.

There are many books on discussion, assertiveness and negotiation skills, see what your library holds.

⇨ TOP TIPS

✓ Being asked to start a discussion is not like being asked to represent your country at football. You are simply 'kicking off'. Make your points clearly and 'pass the ball' promptly. Focus thoughts by putting the main points on a handout or flip chart.

✓ Don't wait for a 'big moment' before contributing. Asking questions gets the conversation going.

✓ Don't be anxious; the quality of your contributions will be super. Say something early on when everyone is nervous and too concerned about his or her contributions to be critical.

✓ Keep discussion points short and simple.

✓ Use examples to illustrate and strengthen your argument. Add references from the start.

✓ Share responsibility for keeping the group talking.

✓ Pre-meeting – meet informally in advance for a short discussion, to kick ideas about and start thinking.

17.2 Discussions online

It is not always possible to get people together. Online discussions (email, chat spaces, message boards, VLE groups or forums, Skype, SharePoint, e-conferencing) can solve teamwork timetabling problems, and are vital for part-time, year abroad and work placement students. They are great practice for business discussions. You have time to think and build research activities into the process. Having started a discussion, you realise that you need additional information. You can find it and feed it in as the discussion progresses. Agree a timescale, possibly:

1. Brainstorm initial ideas, assign research tasks and data collection (Day 1)
2. Research topics and collect data (Days 2–5)
3. Individuals draft their subsections and share them around the group (Days 6–9)
4. Meet to discuss progress, decide on areas that are complete, assess where additional research is needed, assign further research and writing roles and tasks (Day 10)

☺ It was ace having Skype meetings. Three people in one flat, two in theirs and Jan at home. Worked so much better because we could see each other.

5. Individuals redraft sections and circulate them again (Days 10–14)
6. Decide whether we are all happy with this? Pass final version to the editor (Day 15)
7. Team members treble-check the figures and references (Days 16–18)

8. Finalise cover, contents and abstract. Everyone has a final read. Add page numbers. Double-check submission requirements are met (Day 18)
9. Submit (Day 18!)

Some ground rules will help:
1. Agree a date to finish (two weeks), to read contributions every x days (two or three days), and that everyone will make a minimum number of contributions (three).
2. Appoint someone to keep and collate all messages so there is a final record (Joe).
3. Appoint a 'devil's advocate' or 'pot stirrer' to ask awkward questions and chivvy activity (Sandy).

Think flexibly

4. Ask someone to summarise and circulate an overview at the end, especially if there is a group report to create.
5. Be polite. In a conversation you can see and hear when someone is making a joke or ironic comment. The effect is not always the same on screen.
6. Where further research is required, attempt to share tasks evenly.

Replying instantly is generally a good idea; it's what happens in face-to-face discussion and first thoughts are often best. Some people 'lurk' quietly, listening rather than comment-

I only read postings of 4 lines or less ...

ing, which happens in all discussions. One of the more unhelpful actions is someone writing a 3,000-word essay and mailing it to the group. This is like an individual talking continuously for an hour. It puts off the rest of the group who will feel there is little to add. Try to keep contributions short in the first stages. A good way to start is to ask everyone to brainstorm 4–6 points to one person by the end of day two. These are collated and mailed around the group as the starting point for discussion and research.

Modules involving online discussions will have the process set up, usually via the VLE. Some students prefer to have discussions on Facebook and other social networking sites as they 'are more private'. Setting up separate groups on social media sites is fine, but you risk:

☹ missing helpful feedback from the lecturer who could guide the group to think through issues online, just as she would in a face-to-face discussion.
☹ losing whole-class involvement and reducing the diversity of ideas.

17.3 Group management

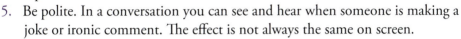

The quality of discussion depends, above all, on the dynamics of the group collaboration (Chapter 16). Some discussions work spontaneously without any problems, others are very sticky. There are no hard and fast rules about behaviour in group discussions, but meetings usually flow well when people take different roles:

☺ Chair (steer) the discussion to keep it to the point, sum up, shut up people who talk too much and involve the reticent.

☺ Note – keep track of the proceedings.

☺ Inject new ideas.

☺ Are critical of ideas.

☺ Play devil's advocate.

☺ Calm tempers.

☺ Add humour.

Take notes during discussions

Formal meetings have designated individuals for the first two tasks (chair and secretary), but in informal discussions, anyone can take these roles at any time. Everyone is better at one or two particular roles in a discussion; what roles do you play and what do you plan to develop? Use **Try This 17.3 NOW.**

⚙ TRY THIS **17.3 – Discussant's role**

Sort the roles people take in discussions into those which are positive and promote discussion, and those which are negative. How might you handle different approaches? (Suggested sort at the end of the chapter.)

Offers factual information	Spots what else – what needs to be followed up
Encourages others to speak	Helps to summarise the discussion
Offers opinions	Asks for examples
Speaks aggressively	Is very defensive
Asks for reactions	Asks for opinions
Diverts the discussion to other topics	Is very competitive
Seeks the sympathy vote	Keeps quiet
Gives examples	Ignores a member's contribution
Summarises and moves discussion to next point	Is (aggressively) confrontational
Keeps arguing for the same idea, although the discussion has moved on	Mucks about

---

Keeping on track

It's very easy for conversations to drift from the main points. Steering the discussion focus to the assignment or module learning outcomes, and the assignment rubric, can be useful.

Asking tutors is not cheating!

17.4 Giving and receiving feedback

Discussions often involve an element of feedback from and to colleagues and tutors. Be constructive because this helps to build everyone's self-confidence. Aim to offer options that encourage the further development of the idea, discussion point or skill. Start with the positive and remember to exemplify your statements, but also remember that if everyone is too nice, you probably won't be motivated to improve. There is an important balance between being positive and proposing changes.

Imagine you are having a team discussion of the first draft of your team report. Some conversational starting points: '*I think this is a useful and concise summary of ..., the section on moraine types has made me understand ..., I am less clear about ...*' is much better than '*I liked it*', and should focus your thinking on revising the section you're '*less clear about*'. Also useful: '*I think it would help me to understand it if we could say a little more about the ...*'. These statements give the discussant a way forward to progressing ideas. Similarly for '*I think I understand but could we add an example of ...*' and '*I think this is a very different way of looking at the issues to the one I*

Discuss everything

had been considering; this is helping me see another side to the argument'. This should prompt discussion about whether you should add or ignore your alternative perspective. Opening up other approaches or options can be very helpful for an individual and group. '*I felt the opening section is Ok so far, but needs more ... (depth, theory, examples, scene setting ...)*'. '*For me, a stronger start, perhaps with some literature cited, would have helped me to jump into the story*', or '*This is a really interesting way of tackling ... What would happen if we looked at it from ...*' or '*Is it worth exploring other angles? For example ...*'.

Receiving feedback can be awkward, many people are not very skilled at giving it. They may be overcritical, intent on getting their ideas across and, in some cases, just plain insensitive. This does not make for easy discussions, so try not to feel defensive and 'got at'. Make sure you understand each feedback point and ask for examples if you are unclear. One useful approach is to note the statements and say, '*Thanks, I will think it over, that's been very helpful*'. If, in the cold light of the next day you feel it was an unjustified comment, ignore it; if it is helpful advice, use it.

17.5 Assessing discussions

In a tutorial or as part of a reflection or learning log activity (Chapter 2) you may be asked to reflect on your contribution to discussion sessions and negotiate a mark with the tutor.

Using **Try This 17.4,** analyse how your colleagues, academic staff, radio and television discussants behave and score against these typical discussion skill attributes. What would you copy or ignore to strengthen your skills?

TRY THIS 17.4 – Assessment of discussion skills

Evaluate a discussant's performance on a 1–5 scale, noting what they do well

Feedback on good points	1 (useless), 3 (average), 5 (brilliant)
Talks in full sentences	
Asks clear, relevant questions	
Describes an event clearly	
Listens and responds to conversation	
Discusses and debates constructively	
Speaks clearly and with expression	
Selects relevant information from listening	
Responds to instructions	
Contributes usefully to discussion	
Reports events in sequence and detail	
Is able to see both sides of the question	
Finds alternative ways of saying the same thing	
Listens to others, and appreciates their input, efforts and needs	

Having analysed what makes a good discussant, have a look at **Try This 17.5.**

TRY THIS 17.5 – Self-assessment of discussion skills

Level 1 students brainstormed the skills they needed to discuss effectively. What would you add? Select 3 items where you want to be more effective and plot a strategy to work on each of these in your next discussion (e.g. I will not butt in; I will ask at least one question …).

Being open-minded	Staying cool
Listening to both sides of the argument	Being tolerant
Using opponents' words against them	Using good evidence
Playing the devil's advocate	Being willing to let others speak
Thinking before speaking	Only one person talking at a time
Summing up every so often	Being firm

Improving

Getting better at discussion and argument needs practice; hearing one's own voice improves one's self-confidence. You can practise in private. Listen to a question on a TV or radio discussion programme. Turn the sound down, take a deep breath to calm down, use it as thinking time. Decide on the first point you want to make. Now say it out loud. Subject matter is not important, have a go. Respond with two points and then a question or observation that throws the topic back to the group or audience. This is a good technique because it shares the discussion with the rest of the audience. Compare your answer with the panellists' responses (use catch-up), remembering to reflect more on the delivery style than the content.

Where points of view or judgements are needed, seek the opinions of people with different academic, social and cultural backgrounds and experience. Their views may be radically different from yours, and change your thinking. Seminars, workshops and tutorial discussions are designed to allow you to share complementary and conflicting views.

A really good discussion will inspire you to research and read further because it opens up new ideas for exploration. To get the most out of a discussion or conversation:

✔ Be positive.
✔ Ask yourself questions like '*How will this help me understand … passenger transport pricing, oscillation ripples, ammonites …?*' and '*What else?*'
✔ Make eye contact with the group.
✔ Give other discussants feedback and support.
✔ Aim to be accurate and on the point.
✔ Include examples and references as you speak.

17.6 References and resources

YouTube has a range of videos with discussions of varying quality. Search: *discussion* or *debate* with *skills, geography, earth science, geology, environment, landscape, management, sustainability …*

Sclafani P, Nygaard C and Thorne C 2017 Applying geomorphological principles and engineering science to develop a phased Sediment Management Plan for Mount St Helens, Washington, *Earth Services Processes and Landforms*, https://doi.org/10.1002/esp.4277 Accessed 16 January 2019.

University of Leicester 2018 Contributing to seminars and tutorials, https://www2.le.ac.uk/offices/ld/resources/study/contributing-seminars-tutorials Accessed 16 January 2019.

University of New South Wales 2018 Guide to Discussion Skills, https://student.unsw.edu.au/discussion-skills Accessed 16 January 2019.

University of Waterloo 2019 Online Discussions: tips for students, https://uwa-terloo.ca/centre-for-teaching-excellence/teaching-resources/teaching-tips/developing-assignments/blended-learning/online-discussions-tips-students   Accessed 16 January 2019.

Quick crossword 2

Answers in Chapter 28.

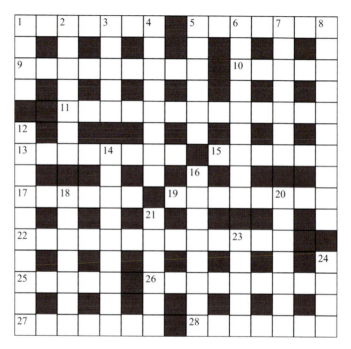

Across

1 Import duties or taxes (7)
5 Ablated (anag.) (7)
9 Fruitfulness, richness (9)
10 Semi-aquatic rodent, *Myocastor* (5)
11 Investigates long-term weather (13)
13 Columbia, Athabasca, Dovre, Campo de Hielo Norte (3,5)
15 In line (6)
17 Early film (6)
19 Dispersion, migration (8)
22 Agreed, sanctioned, signed off (6,7)
25 Main blood vessel (5)
26 Makes acid (9)
27 Weakened, let down (7)
28 Stinking chamomile, made yew (anag.) (7)

Down

1 Porous limestone (4)
2 Reprocess, salvage (7)
3 Moulds, mushrooms and toadstools (5)
4 breathing hole, calipers (anag.) (8)
5 Decomposes timber (3,3)
6 Earth movement science (9)
7 By the sea, easy bid (anag.) (7)
8 Or a tequila (anag.)(10)
12 Neighbourhood (10)
14 Nomadic traveller (9)
16 Ti 22 (8)
18 Not labour or conservative (7)
20 … dancing, model it (anag.) (3,4)
21 Australia, Madagascar and Man (6)
23 … council, outhouse (5)
24 Lattice, network, web (4)

Answers: Try This 17.1 – Working with a brainstormed list

There are many ways to order this information; the headings here are suggestions. Notice that there is some repetition, a mix of facts and opinions. Some points fit under more than one heading.

1. *Slope instability features:* Movement of handrails. Tennis courts on old levelled slip. Evidence that paths across the slips are regularly re-laid, cracked tarmac, fences reset. Promenade held up by an assortment of stilts, walls and cantilever structures where slope below has been undercut. Castle fortification, some walls now undermined and eroded. Path covered by slumping clay. Vegetation eroded from wetter clay slopes.

2. *Geology and geomorphology:* Complex geology, land slips, big fault structures. Undercutting of cliffs by waves. Rocks at cliff foot dissipate wave energy. Soil overlying clay.

3. *Slope repair and restoration features*: Concreting the slopes prevents soil erosion. Fine and coarse netting, geotextiles, used on clay slope to help stabilise the soil. Bolts into the rock face to increase stability. Rock armour installed at the foot of recent landslip. Wall to prevent boulders from cliff hitting the road. Tree planting on recent and older slip sites. Drainage holes in retaining walls to reduce soil water pressure.

4. *Tourism factors:* Café for tourists and visitors at foot of slip area. Hotel, amusement arcade and tourist shops – noisy and unsightly. Harbour has mix of marina, fishing and tourist fairground rides. Victorian swimming pool, in need of repair.

5. *Transport issues:* Car parking demands, unsightly in long views. Car pollution. Car parking unscenic. Caravan park on cliff.

6. *Beach issues:* Sea wall prevents undercutting. Rocks at cliff foot dissipate wave energy. Groynes prevent longshore drift. Repairs to sea wall.

7. *Other issues:* Tourist facilities destroy foreshore view. Soil overlying clay rock. Vegetation worn on paths and alongside paths. Litter needs collecting from beach. Road built over old stream, presumed piped underground. Varied quality of coastal path, not always well signposted.

Answers: Try This 17.3 – Discussant's role

Positive points: Offers factual information; Spots what else – what needs to be followed up; Asks for examples; Encourages others to speak; Asks for reactions; Offers opinions; Helps to summarise the discussion; Asks for opinions; Gives examples; Summarises and moves discussion to the next point;

Negative points: Seeks the sympathy vote; Is very competitive; Speaks aggressively; Is very defensive; Keeps quiet; Ignores a member's contribution; Mucks about; Diverts the discussion to other topics; Keeps arguing for the same idea although the discussion has moved on; Is (aggressively) confrontational.

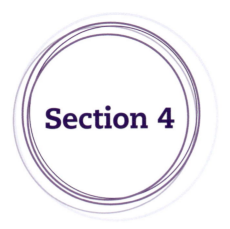

Assignments and assessment

In this section the focus is on the different ways in which you will be assessed, from essay writing to dissertations. Your higher education challenge is to become reasonably expert as a geographer, environmental scientist or geologist so that the discipline becomes part of your life. Good assessments are designed to give the opportunity to learn and be creative in presenting your arguments. The skills you're developing are in evaluating the diversity of methods, philosophies, case examples and theories in order to weave a selection of these ideas into your own coherent story, and that takes time and thought. Every assignment you present will be unique, because it's based on what you've done, found, discussed, read and how you've made sense of your information and data. Chapters 5.9 (Writing), 5.10 (Word processing) and 5.11 (Acronyms) have useful prompts for the writing process.

As you reflect on a case study, reading or class session try to distinguish between facts (which you need) and the main concepts or theories. It's the **active** process of thinking and talking about these ideas and theories, making notes, testing out ideas in discussions, which will give you the framework to make sense of isolated facts, and you should remember them for longer. Using a technique on the field-class day gives you experience of the process, some deeper thinking later involves working out where else that technique would be valuable, and why you would choose this technique rather than any other. It's that further thinking that leads to higher-quality assignments.

Applying a three-phase approach to assignments and projects allows you and your team time to evaluate, reconsider and refine your thinking. It also provides reasons to celebrate at the end of each of the three stages. How do your actions mesh with this approach:

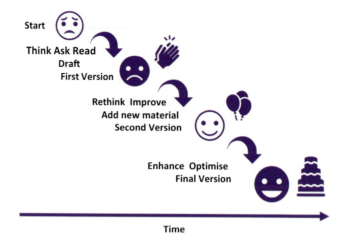

18

Effective essay skills

I'm getting on with the essay, I decided the page numbers go bottom left, they look really neat. Writing next …

Many of the chapters in this book start with some reasons why practising a particular skill might be a moderately good idea, and an analogy involving practising playing musical instruments or sport. Now demonstrate your advanced creative skills by crafting two well-argued sentences using the following words or phrases: reports communication language persuasion cheerful examiners lifelong clarity with needs solos practise bagpipe.

This jumble of words may give you an insight into how annoying examiners find disjointed, half-written paragraphs, with odd words and phrases rather than a structured argument. Such paragraphs have no place in essays. If this is the first chapter you have turned to, keep reading!

What is your experience of the practice of writing (Chapter 5.9), and writing essays? At university, an essay is usually 1,000-plus words of well-argued writing that demonstrates you can research. This chapter gives you the important points for writing for geography, earth or environmental sciences

> *I did History for A level, I understand grammar, my essays are OK, why do tutors keep assuming students can't write?*

essays. BUT there are many books and websites that discuss writing skills in more detail. Mature students who have '*not written an essay in years*' or those who did science A levels and '*haven't written an essay since …*' should start with this chapter and check various websites – see the resources section (18.8) for starting points.

> *Last time I wrote an essay I was 12. What is an essay? What do tutors want?*

Tutors are often asked '*what a good answer looks like*'. The most helpful response involves comparing and evaluating different pieces of writing. This chapter includes some examples of student writing at different standards. You are asked to compare the extracts to get a feel for the type of writing you are expected to produce. Space restricts the selection, but you could continue by comparing essays with friends, study buddies and members of your tutorial group.

The start of every piece of writing is really important, as there's no second chance to make a good first impression, and it puts the tutor marking it into a good mood. And the very end is also important – soon the marker will be deciding the grade!

All essays need a balance of personally researched and lecture-based evidence. This wealth of material needs sorting out with an initial plan, then some rethinking, writing, further research and rewriting. Then a heavy editing session, where the initial long sentences are cut down to shorter ones and paragraphs are broken up, so that each paragraph makes a separate idea, thought or point. The first version of anything you write is a draft, a rough-and-ready first attempt, requiring development and polish into the second and third versions before it is a quality product. A lot of possible marks disappear because the first draft is submitted as the final product.

Recognise that writing evolves and that exploring ideas is never simple. It takes time. There will always be a range of ideas, layers of complexities and examples. This means that as you research and write:

- new data and information will appear and change your ideas;
- early reading matter may need to be excluded;
- new arguments will need to be woven into the text;
- early ideas will evolve and may need to be restated in a new way;
- the examples carry the arguments forward; and
- there is a good balance of space given to the big ideas relative to the smaller details.

Really good essays evolve over a month of thinking, researching, drafting and revising. Plan your time (Chapter 3), and remember that Chapters 12, 14 and 15 have lots of useful advice.

18.1 What kinds of essays are there?

The 'tell me what you know about …' style of essay should be disappearing from your life. University questions usually require you to think about information independently to create a persuasive argument. Questions will ask you to analyse, criticise, examine and debate ideas in a structured way, using apt examples to illustrate your arguments. Interweave lecture material with personal research findings and ideas for high marks. Reproducing the facts and arguments from a lecture may get a mark of 30–50 per cent. Painful but true. For 50 per cent plus you must show the examiner that you have thought through the issues, sorted out what it all means, and put the argument in your words (Figure 18.1). OK, that's my opinion. Ask your tutors what they expect.

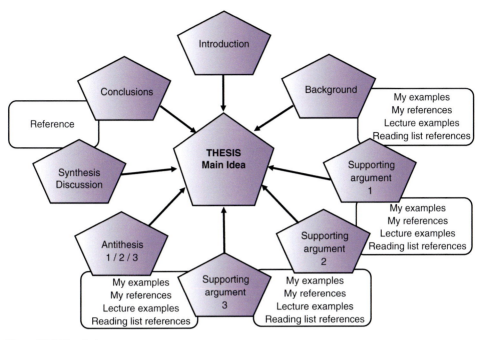

Figure 18.1 Structuring an essay

You will increasingly be asked for discussion rather than descriptive essays. Compare these questions:
1. *Descriptive.* Outline and justify the methods of store location you would recommend to a retailer.
2. *Discussion.* Store location is a retailer's minefield. Discuss.
3. *Descriptive.* Describe the role of potassium argon dating in geology.
4. *Discussion.* Evaluate potassium argon dating as a tool for geologists.

The descriptive essay title has pointers to the structure of the answer. The discussion essay needs more thought and planning; you must establish your structure, and write an introduction to signpost this for the reader. The introduction leads to a series of linked arguments supported by evidence (*references*), leading to a conclusion that follows from the points you have made. Question 3 lets you explain one dating method, question 4 requires information on the alternatives to potassium argon dating so that you come to evidenced conclusions about its value. Including material that veers off at a tangent or is irrelevant, or presenting evidence in ways that do not really support your case, will lose marks. Archaeologists use potassium argon dating. There are archaeology references and case studies that may be useful or be too off the point for this essay – you decide.

What are the supporting ideas?
Most essays explore issues with no right answers. You consider the various dilemmas, debates and decisions that are the essence of real situations, provide evidence and reach a balanced conclusion. Sample questions: 'Analyse the causes of

> *My lecturer really means it when he says he wants to be surprised by the answers. He wants 1500 words, references, short paragraphs but the rest is up to me. Scary stuff.*

evolutionary mass extinctions', 'Compare the strategies homeless men and women develop to survive in urban areas', 'Creationism presents a valuable counterbalance to science. Discuss', 'Consider whether global climate modelling presents an impossible challenge', 'There should be no more nuclear power stations', 'Will focusing environmental and health policies on minority and low-income communities lead to environmental justice?', and 'What is the evidence to support the argument that ecological niches will always heal themselves?'. All of these are contentious.

18.2 Analyse essay titles carefully

Take time to analyse the question; it provides a structure for your answer. For example:

Discuss the causes, effects and methods to reduce hazards associated with earthquakes.

> *It was sooo stressful, I had no idea what was wanted, trying to fit into this new way of thinking, working. The draft essay we did in first term was v. v. v. useful, not so much mine, but looking at the rest of what my tutorial group did. Between us we shared so much useful feedback.*

Anyone who thinks that this is an essay about earthquakes is in serious trouble. It is an essay about fire, flooding, landslides, tsunami, collapsing of bridges and disruption to communications: a hazards essay. Earthquakes get a passing reference in exemplification, as for example, 'The collapse of housing stock in the Turkish earthquakes of …' or 'The severing of network links where roads collapsed during the San Francisco earthquake led to …'; they are not the essay's focus. Equal weight needs to be given to the three main sections of the essay – 'causes', 'effects' and 'methods to reduce', with each aspect discussed in relation to each hazard.

Describe and critically evaluate the factors that are considered to drive glacial cycles.

What are the key words? The problem with this title is that there is too much potential information to discuss: the influence of plate tectonics, orogeny, the raising of the Tibet plateau and consequent feedbacks to climate, changing global CO_2 and temperature balances, sea temperatures and currents, volcanic dust, astronomical theories (including Milankovitch, equinoctial precession, orbital eccentricity, sunspot and solar flare theories). There is an excess of information and data. Most students attempt to include everything they know and thereby throw away the 50 per cent of the marks for the 'critically evaluate' section.

Now look at **Try This 18.1**.

 TOP TIP

✓ Assume all essay questions include the phrase '… with reference to specific geographical/environmental/geological examples'.

TRY THIS 18.1 – Key words in essay questions

What are the key words and potential pitfalls with these titles? Some thoughts are at the end of the chapter.

1. Discuss the role and responsibilities of science and scientists for society.

2. Communications, internal commerce and energy are the sectors that are usually identified as the most serious 'bottlenecks' in contemporary Chinese development. Explain the weaknesses in one of these sectors.

3. To what extent are growth and change impeded by archaic social structures in either Latin America or a selected country in Latin America?

4. Critically explain how Andrews (1983) derived his bedload entrainment function.

5. Explore the arguments that further drilling for oil should be permitted in Alaska.

6. Evaluate the role of mass media reportage of environmental issues.

18.3 Effective introductions

A good introduction serves two purposes. It outlines the general background and signposts the structure and arguments that follow. It gives the reader confidence that you are in command of the topic. Short is good, 100–200 words, roughly half a side, is usually plenty. If the introduction is longer consider moving some material to later sections.

Introductions require thought. Good opening paragraphs are often written last. Planning time is vital. A provocative headline opening, grasping the imagination of the reader is a good wheeze, BUT it must be integral to the essay. Lengthy case material and examples are not usually good opening material. If there is an example in the first paragraph, limit it to one sentence – further details belong in the body of the essay. Here is a reasonably good introduction (it could be improved):

Explain, with examples, why river flows in urban channels are rarely natural.

Urbanisation and the consequent paving of natural surfaces with cement, roofs and tarmac has changed the natural water balance within cities: there is less infiltration, less soil storage and increased rapid surface runoff (Wellis and Mack 2030). Urban flooding can be both personally inconvenient and economically disruptive. Engineering works have channelled runoff efficiently through drains and sewers to minimise surface inundation. The demand for water in the city is high, and in developed countries particularly, 'this demand has increased through the twentieth century as lifestyles have changed' (Gluggit 2030: 45). River water is extracted for drinking from local sources where possible, reducing flows. Effluent is normally treated within the catchment and returned to the nearest river, adding to flows. The competing demands for extraction and disposal of water, combined with maintaining minimum flows in drought periods and controlling flood flows mean that urban channel flows are managed and regulated rather than natural. These demands are not always compatible; a strategic planning issue for water managers. A range of case studies from the UK and USA illustrate the issues and conflicts that arise.

This introduction summarises some of the major interlocking issues, indicates that there are no simple answers and that examples will be taken from two developed countries. The Gluggit direct quote is in '…' (✔) and is cited with the page number (✔), but is a not particularly helpful statement in this example. Use quotes to make strong points.

This next example contains a series of true statements, but gives no indication of the structure and content of the essay. AVOID THIS approach please.

Discuss the economic impact of floods in urban areas.

Over history, many very important urban developments have grown up besides or around fluvial channels, which are susceptible to flooding. Noah was one of the first to avoid a flood. Over this same period, the inhabitants of those urban areas by rivers have tried to manipulate the natural channel so as to reduce the risk of flooding and thus economic impact on that area. Although flooding may take place at any point along the channel's path, it is in urban areas that the greatest density of housing is affected, thus maximum disruption takes place and the largest economic impact is felt.

The first sentence is long and woolly; it could read, 'Historically, flood-prone riverside sites have been chosen for settlement' (11 words, not 20). Similarly, the start of sentence four could read, 'Although flooding may occur anywhere along the channel' (8 words rather than 12). The second sentence is totally irrelevant and unconnected. There is no indication in this paragraph of the topics the essay will address, no mention of direct and indirect costs, insurance and liability, planning legislation to control floodplain development, modelling insights from hydrological and economic aspects and case examples.

A more headline style may not always be appropriate, but might start:

> *Who pays if your business is flooded by storms, snowmelt, hurricanes, water backed up from overloaded storm sewers, or the failure of a dam or embankment? More than 90 per cent of the disastrous floods in the US are not declared as federal emergencies. Flooded properties do not qualify for financial assistance. Many UK residents do not have flood insurance. If your business is flooded, the costs are your problem.*

Or,

> *You do not have to live on a floodplain to be flooded in an urban area. Flooding following unusually large storms, the backing up of sewers, rapid snowmelt, collapse of floodwalls, dams or embankments, or the failure of a mains water supply pipeline can hit any home or business.*

Pretend to be an assessor, take a critical look at **Try This 18.2.**

TRY THIS 18.2 – Evaluate an introduction

Pretend to be an examiner looking at these four introductory paragraphs to the same essay. You need no information and data on Canada or Wetlands; a gap in your content knowledge helps you concentrate on the main issues. Questions to ask yourself include:
- Is the general case outlined and explained?
- Does the introduction indicate how the author will tackle the essay?
- Is the language suitably technical (or 'grown-up', as a tutee once described it)?

The essay title is horribly general, which makes a good opening particularly important. Once you have had a think, compare your thoughts with the comments at the end of the chapter.

Discuss human impact on Canadian wetland landscapes in the last 100 years.

Version 1 Wetlands are described as transition zones between terrestrial landforms and water bodies. They resemble both uplands, due to their ability to support emergent plants, and aquatic regions, because of the domination of the areas by water. Even though these highly productive areas contribute 14 per cent of Canada's total land surface, they were not considered initially to have much value because they made transport difficult across Canada, whilst at the same time harbouring insects that led to many deadly diseases.

Version 2 Wetlands perform a range of valuable functions, including reducing flooding, filtering polluted waters, defending shorelines and as habitats for many species, including migrant birds, which makes them landscapes worth retaining. They are found wherever the water table is at or near the surface for most of the year and include salt and freshwater marshes, bogs, sloughs, fens and the margins of lakes and rivers. They are dependent for their structure and function on a sustained hydrological and nutrient status, but this also makes them vulnerable. A change in the water supply through drainage, drought or flooding will upset the ecology, as will a sudden increase or decrease in nutrient supply (Verboggi 2030). Recognition of the value of wetlands is a recent phenomenon. In the past they were considered as wastelands, dangerous quagmires that could trap animals and the habitat for insects that cause diseases such as malaria. Early settlers in Canada could afford to ignore wetlands – there were more promising lands to farm – but as the pressure on land grew through the twentieth century, wetlands were increasingly reclaimed. In the past 100 years a diversity of competing interests has led to their reclamation for agriculture, drainage schemes, urban expansion, forestry, and peat extraction for horticulture and fuel (Albertus and Brit-Columb 2030). The competing issues and impacts are illustrated here through a range of case examples, from Arctic to continental wetlands, highlighting the vulnerability of the sites and the problems of protection and reclamation. However, despite the loss of 70 per cent of Canada's wetlands, there are good examples of conservation and site protection, exemplified through RAMSAR (2018) and other sites.

Version 3 Due to the ever-increasing population and rapidity of land use change, the prevalence of wetland areas (wetlands are a transition between landforms and water bodies) is depleting. Within urban areas wetlands are often the last to be developed, and as cities have expanded through this century there has been a rush to fill these in too. It is man's impact that, particularly through the last century, has influenced them most. Many have been depleted due to their site near to main rivers, coasts and bays, and therefore have been developed due to their necessary use for transportation. The conversion of lakes and reservoirs has added to this. This essay will examine some of the impacts on wetlands in the last 100 years.

Version 4 'Canadian wetlands are extensive but vulnerable to exploitation and reclamation' (Environment Canada 2030: 123). The pressures of agriculture, drainage schemes, urban development, forestry and peat harvesting have significantly reduced their extent and this has had an impact on the wildlife which use wetlands as permanent or migration habitats. This essay will assess the extent and processes of change, using examples from wetlands across Canada, and then focus on the particular issues raised by the Copetown Bog near Ontario. As this case reveals, there are grounds for optimism as the significance of wetlands (RAMSAR 2018), and the need to conserve them, is nationally and internationally recognised.

18.4 The middle bit!

This is where to put information, data and commentary, developed in a logical order like a story. You need a plan. Creating a list of bullet points, putting them in order and writing a paragraph for each is a great plan. Subheadings make plans clearer to you and the reader. Whether you use subheadings mentally, or mentally and physically, is a personal matter. Ask your tutor if s/he minds subheadings in essays. Most writing involves changes in the order of material as your understanding and argument develop.

Examples are the vital evidence that support the argument. Where you have a general question like 'Describe the impact of satellite TV on global marketing', aim to include a range of case examples, BUT if the question is specific, 'Comment on the impact of Sainsbury's marketing strategy', then the focus is primarily on Sainsbury's. In the 'global marketing and TV' essay, remember to make the examples global. There is a temptation to answer questions like this with evidence from the USA, Europe and Japan, forgetting the developing world. Use as many relevant examples as possible, with appropriate citations. Generally, many examples described briefly get more marks than one example retold in great depth. Examples mentioned in lectures are used by 80 per cent of your colleagues. Exhilarate your examiner with a new example (*more marks*).

Graphs, figures, and pictures

Graphs, figures, photos and pictures illustrate and support arguments in fewer words than a description of the same material. Insert, label them fully and accurately and refer to them in the text. If there is no instruction to 'see Figure 1', the reader won't look at it (*no marks*). Clearly labelled figures can then be cross-referenced from anywhere in the essay.

Reference appropriately

All sources and quotations must be fully referenced (see Chapter 14), including the sources of diagrams, photos and data. Avoid citing the obvious, '... *water runs downhill (Noah 2350BC)*'. Aim for technical content. '*Alkimim and Clarke (2018) show that the impacts on greenhouse gases from producing ethanol from sugar cane depend on the production method*'. This statement is true, but from the same paper you could write:

> '*Calculating the carbon debt when different types of forest in Brazil are cleared for sugarcane production yields results "equivalent to 608 Mg CO$_2$ ha^{-1} for the Amazon, 142 Mg CO$_2$ ha^{-1} for the Cerrado and 212 Mg CO$_2$ ha^{-1} for the Atlantic Forest with respective payback times of 62, 15 and 22 years" (Alkimim and Clarke 2018: 66). They calculate that planting in pastureland rather than forest would reduce the carbon debt, making ethanol production from the sugar cane more efficient and reduce greenhouse gas emissions*'.

This is a more detailed, more technical example, including a direct quote from the paper so the page number, 66, is included.

18.5 Synthesis and conclusion

A conclusion is the place you have reached when you are tired of thinking.

Synthesis

The discussion or synthesis section allows you to demonstrate your skill in drawing together the threads of the essay. You might express the main points in single sentences with supporting references, but this is just one suggestion. Here is an example of a synthesis paragraph. In answer to the essay question 'Evaluate the principal theories of infiltration', paragraphs 2–8 outlined the main theories with the equations, providing the information section of the answer. This was followed by:

> *Current understanding of the processes of infiltration is influenced by the three main theories outlined above. For dry and sandy soils with high permeability, the Green and Ampt (1921) model is appropriate. The Phillips equations (1937) have been widely and successfully used, for example Engman and Rogowski (1976), in an overland flow routing model. In saturated and semi-flooded soils, Hillel's (1986) model provides a more accurate set of results than the Phillips approach. This is because permanent or semi-permanent ponding on the surface reduces the air infiltration element from the equation. Neither of these models is appropriate in permafrost areas with frozen soils. For these environments, the results of Dunne (1964), although limited to soils in Nebraska, seem to provide a more accurate model. In addition to these field approaches, the data from laboratory experiments by Jackson (1967) and Burt (1982) have improved our process understanding in detail. However, these approaches, although apparently successful, have yet to be tested at the field scale. At present there are a suite of approaches available to the hydrologist. However, s/he needs to select the technique that is most compatible with prevailing environmental conditions.*

So what do you think? Lots of information, interwoven references, useful information – this should get high marks and is a style to aim for. The last three sentences are very good. If the facts were right it would get high marks. BUT this particular paragraph will be lucky to score 40/100. The authors are all well-known hydrologists and some did the work referred to, some did not and all the years are wrong. What is worse is that the most recent reference is 1986 – has nothing happened since then? This is an unacceptably dated answer for 2019+. Examiners notice spoof and imaginary references.

Conclusions

End with a flourish, not a yawn! The conclusion should sum up the arguments. Do not repeat statements from the introduction or introduce new material. This causes markers to comment: 'The conclusions had little to do with the text'. Many markers will jump straight from the starting paragraphs to the conclusion to see where the piece is heading. Before submission read the first and last sections together and make sure they link up.

It is hard to judge the standard of conclusions in isolation, because the nature and style of a conclusion depend on the content of the essay, but it helps to look at other examples. I advocate reading articles in *The Economist, Nature, New Scientist* and *Time*. There is something relevant in most issues. Despite the caveat above, criticise the examples in **Try This 18.3.**

TRY THIS 18.3 – Good concluding paragraphs

Read the following concluding paragraphs and consider their relative merits. Have a think, then decide whether you agree with the feedback at the end of the chapter.

Version 1 It is therefore possible to see that man has had a mega impact on the landscape as the land devoted to agriculture quadrupled and there was an explosion in urban development. This impact was caused by the discovery that wetlands have many important values and perform many worthwhile functions. The components of the system are seen to be interconnected, and destruction of one part of the system will have effects, probably adverse, elsewhere in the system. This has meant that for the last 25 years there has been less damage, but only a great deal of time and conservation will reverse the extraordinary exploitation that has occurred on Canada's wetland landscapes.

Version 2 All in all, there has been a lot of pressure on Canadian Wetlands over the past hundred years due to man's impact, but things look set to improve, with talk and actions of restoring many wetlands areas, and also new international regulations and laws for their protection.

Version 3 In addition to the impacts I have considered, threats likely to continue or increase include the harvesting of black spruce (*Picea mariana*), use of wetlands for wastewater treatment, burning, which may lead to the loss of native species and the invasion of Eurasian weed species, and increased accumulation of sulphur and heavy metals from acid rain. The impact of man has led to the loss, or serious degradation, of these vulnerable ecosystems, which are effectively a non-renewable resource. The recreation and educational value of wetlands has been discussed. There is a dilemma for managers here, who need to protect the ecology of sites while giving appropriate access for eco-tourists, hikers, ornithologists, fishermen and hunters. Good *et al.* (1978) pointed out that much wetland management derived from common sense rather than science. In 1999 the science base has increased, the complexities of ecosystems interactions are better understood. The RAMSAR Convention (2011) has provided valuable guidelines for wetland preservation, but there is still much to be done to prevent further wetland destruction.

Version 4 Canadian wetlands were drastically altered by the intervention of humans. Seventy per cent have been reclaimed for a variety of purposes, including agriculture, urban development, industry, energy development and harvesting. There is a consequent loss of habitat and reduced species diversity. Recent changes in human use of wetlands have led to some abandoned agricultural sites being returned to wetland, but agriculture is still the dominant activity.

Version 5 To conclude, I think it important to highlight the damage done by man to wetlands, both on a national scale and at a local scale where only small amounts of interference can cause potentially devastating effects. This essay has shown that without greater management of these environmentally sensitive areas, man will not only destroy a vast educational resource but also affect the entire global atmospheric system.

18.6 Assessment

Either use your department's essay feedback form or Figure 18.2 to critically self-assess your first and second drafts. The results indicate where to focus your next research and writing efforts.

Essay Title ....

Good points!	1 >70	2.1 60– 69	2.2 50– 59	3 40– 49	Fail <39	Avoid!
Information 30%: topic covered in depth; multiple examples; embedded references						Superficial content; few examples; few references
Argument 40%: logical presentation of ideas; one main idea per paragraph; good use of illustrations – diagrams, graphs …						Disorganised ramble; few illustrations
Creativity 20%: includes some new element – idea, presentation of data, graph …						No evidence of personal thinking about the ideas
Presentation 10%: correct grammar and spelling throughout; clear layout; quotations in '…', complete and accurate reference list						No or few references; spelling and grammar errors

Figure 18.2 Self-assess an essay. (Note the percentages in column 1 are guides which will vary between assignments.)

Another way of developing evaluation skills is with **Try This 18.4**, which builds on an idea from note-making by using codes to locate the different sections of an essay and compare their relative weights. **Try This 18.5** suggests structure by considering the mark distribution. If the balance of the reward changes, then so does the content.

> **TRY THIS 18.4 – Analysing essay structure**
>
> Analyse an essay you have written recently using coloured pens, or a code system.
> Mark in the margin the sections which show Information and data, Analysis, Synthesis,
> Evaluation, and Creative abilities. Now consider the balance of these five sections. Where
> could the balance of the essay be altered to improve it? Critical questions are:
>
> **?** How would you rewrite this essay to increase the analysis section?
> **?** How would you redesign to increase the evaluation content?
> **?** Look at the structure of your essay:
>   ✓ Does the argument flow logically?
>   ✓ Is there a good opening paragraph?
>   ✓ Are the examples relevant?
>   ✓ Are the arguments summarised effectively?
>   ✓ Are the conclusions justified?

> **TRY THIS 18.5 – Marks for what?**
>
> Take any essay title (the last or next) and work out a plan for researching and answering
> the question when the mark distribution will give credit in different ways:
>
> 1. Information and data 20%; Analysis 20%; Synthesis 20%; Evaluation 20%;
>   Professionalism 20%; AND
>
> 2. Information and data 10%; Analysis 50%; Synthesis 20%; Evaluation 10%; Creativity
>   10%; AND
>
> 3. Information and data 30%; Evaluation 70%.

Getting advice

Showing a draft essay, or any document, to someone else for comment is not cheating. It is normal practice in business and academic publishing, as shown by the acknowledgements at the end of many published papers where there are thanks for reviewers' feedback. You do not have to take note of the comments, but having an independent check on grammar, spelling and someone asking awkward questions does no harm. Do it early so that you can revise your essay.

Always number all
the pages

18.7 Getting the English ~~right~~ better

Try to NEVER cross ~~nothing~~ anything out! Keep sentences short wherever possible. Practise with **Try This 18.6.**

TRY THIS 18.6 – Shorten these sentences

Answers at the end of the chapter.

Wordy	Improved
In many cases, the tourists were overcharged.	
Microbes are an important factor in soil processes.	
It is rarely the case that sampling is too detailed.	
The headman was the proud possessor of much of the land in the vicinity of the village.	
Moving to another phase of the project …	
Chi-square is a kind of statistical test.	
One of the best ways of tackling prison reform is …	
The investigation of cross bedding at Ramsgate continues along the lines outlined.	
The nature of the problem …	
Temperature is increasingly important in influencing the rate of snowmelt.	
One prominent feature of the landscape was the narrow valleys.	
It is sort of understood that …	
It is difficult enough to learn about stratigraphy without time constraints adding to the pressure.	
The body of evidence is in favour of …	

Technical terms

Good professional writing in every discipline uses technical language, defining technical terms when necessary. Definitions can be very important: housing policies in Leicester are rather different from those in Sao Paulo; an intense rainstorm in Singapore is considerably more intense than one in Glasgow; a socialist, liberal or conservative government means different things under different regimes.

There is no need to define technical terms that are in everyday use and where you are using words in their normal sense; assume the reader is intelligent and well educated. In an essay on hydrological flows in river basins, paragraph 2 started with: '*Water covers a third of the earth's surface, mostly as oceans. Oceans are expanses of saline water which vary in dimensions, depth and width, and form e.g.: calm, rough, warm, cold*'. In my opinion this student wasted time on the second sentence – an examiner knows what an ocean is. Keep the language suitably technical and the sentences simple. Use the correct technical terms wherever possible and avoid being unnecessarily long-winded, as in '*snowmelt is largely a seasonal phenomenon, which occurs when temperatures rise after frozen precipitation has fallen*'.

Startling imagery

'*It is a truth, universally acknowledged, that a man in possession of a large-screen satellite TV on Cup Final Day, must be in want of a six-pack*'. The impact is greater for being the antithesis of Jane Austen. However, this comment, while arresting, is dangerously stereotypical, far from politically correct and gendered. Creative use of language is great; please avoid the temptation to bowdlerise or tabloidise inappropriately. (Search for 'Thomas Bowdler 1754–1825' for the origin of bowdlerise.)

One idea per paragraph

In a lengthy essay, restricting paragraphs to one idea plus its supporting argument should make your message clearer. Use **Try This 18.7** to analyse one of your essays.

> ⚙ TRY THIS **18.7 – One idea per paragraph**
>
> Reread your last essay underlining the ideas and supporting statements. If each paragraph has a separate idea and evidence, award yourself a chocolate bar and cheer. If not, redesign a couple of paragraphs to disentangle the arguments and evidence. The challenge is to make the evidence clear to the reader by separating out the different strands of the argument.
>
> Note: This is a good activity with the first draft of every essay, report …

Synonyms

Part of the richness of English comes from the many synonyms that add variety, depth and readability to writing. Take care you write what you mean to write! A common error is confusing *infer* and *imply* – they are not synonyms; *infer* is used when drawing a conclusion from data or other information, *imply* means to suggest or indicate. We might infer from a dataset that the world is flat, or a lecturer might imply during a discussion that the world is flat. Practise with **Try This 18.8.** Look at any sentence you have written, and play around with the thesaurus settings to find synonyms you might use. If you tend to overuse certain words, make a list and substitute some of them WHERE APPLICABLE.

TRY THIS 18.8 – Synonyms

Which of the two synonyms makes sense in their geographical, environmental or geological context? Answers at the end of the chapter.

1. Studies on litter and organic matter dynamics also assist awareness/understanding of nutrient mineralisation.

2. My research project uses focus group interviews as its main/paramount data source.

3. The discrimination/distinction and mapping of rock types are a major focus of remote sensing in geology.

4. River networks can be deduced/extracted from satellite images, allowing controls on drainage patterns to be analysed.

5. What economic logic elucidates/explains plant closings and firms' exit from industries?

6. Modelling is particularly important if real appraisals/estimates of the extent and timing of hazard events are to be of practical value.

7. Past experience/practice of negotiating emission agreements suggests that, despite legislation, success in emission control is by no means assured.

8. Any process by which a marine-ice sheet may form must overcome the problem of enhanced calving rates associated/comparable with deeper water.

9. There is some limited documentation of the effects of channel and water management on the morphology/shape of channels in the Mediterranean region.

Argument by analogy

Analogies can be very useful, but this example shows they can get out of hand: '*A fast breeder reactor is more efficient than a normal nuclear reactor. It can be thought of as a pig, eating anything, rather than a selective racehorse*'. The first sentence would be better standing alone. '*An unconformity is where two beds of rocks are out of order, like a sponge cake with the jam missing*'. This last analogy might be fine during a conversation in a quarry, very helpful in getting novice geologists to understand, but is out of place in an essay. Be technical.

I know what I meant!

Tutors struggle to encourage students to reread and correct written work. Rereading reveals illogical statements and where crucial words are missing. Here, for your amusement, are some essay statements that do not exactly convey the writer's original intention: '*For many centuries cars have been causing problems on the roads*'. (When were cars invented?) '*No data can be gathered as the area is relatively flat*'. '*Trace fossils are thought to have existed as early as very early on*'. '*Large-scale windmills should be located in areas where winds blow almost inconsistently*'. '*Laterites grow in topical environments*'. '*... when it's really dry the soil gets to wilting point* *and the plants become extinct'. 'Scrapping rocks shows the colour of the minerals. Iron is red when scraped'.* (Scraping not scrapping.) *'A random model has lots of strange processes usually found by throwing dice. To get a weighted random model you need a weighted dice'. 'Blackpool has made concreted efforts to manage tourists'.* (Try concerted instead.)

Abbreviations, acronyms and txtg

Use well-known and established acronyms (Chapter 5.11), but replacing long words or phrases with initials or abbreviations is lazy. The first time a phrase is used put the abbreviation immediately afterwards in brackets.

> *No text speak please. Der r no occasions whr wrds hweur specl cud tel yr story, or wil gt U mrks. Njoy txtng @ oder tyms WFs.*

Using nouns as verbs

Please avoid transforming nouns to verbs as in 'to podium' or 'to novelise', or verbs to nouns. Avoid 'the rurality of the site' (use the rural site) or 'the malnutrite stock' (the stock suffering from malnutrition) and similar phrases. Creating new words is not good.

Colloquial usage

Please avoid regional or colloquial terms which may not be understood. Writing as you speak is also a trap, as in these student essay examples: *'He therefore put to greater emphasis on the results of …'* It should read, 'He therefore put too great an emphasis on …' And '… *large events will of always permeated through …*', should be '… large events will always penetrate through …'

> Please don't use *cool, wicked, maybe, neat, so dead,* and *bingo* as descriptions.

For more examples, search: 'commonly misused English words'.

Spelling and punctuation

Spelling is a potential minefield: use the spell checker, and then check carefully for errors it missed; the following words are spelt 'correctly': *'the Haddock shopping centre'* (the example was Haydock), *'models can be said to be lumped or disintegrated', 'Fare Trade', 'focused on the shear pace of change', '… from the census we calculate the morality of inhabitants', 'middlemen can charge exuberant prices', 'supper continents', 'the competition between predators and pray',* and *'asses'* which so often appears rather than 'assess'.

The excessive use of exclamation marks!!!!!!!!!!!, is not good! And please never start a sentence with And, But or BUT. There are plenty of examples of its (strictly) incorrect usage in this book, where BUT and capitalisation are used to emphasise points.

> ⇨ **TOP TIPS**
>
> → Make time at the end of an assignment to reread and redraft, correct spelling, insert missing words, check grammar, tidy up diagrams and insert references.
> → Check that your arguments are logical.
> → Read what is written, not what you meant to write.
> → Keep sentences short and TO THE POINT.
> → Ensure paragraphs address one point only.
> → Be consistent in use of fonts and font sizes, symbols, heading titles and position, bullet points and referencing. Decide on your style and stick to it.

18.8 References and resources

Alkimim A and Clarke KC 2018 Land use change and the carbon debt for sugarcane ethanol production in Brazil, *Land Use Policy*, 72, 66–73.

CCC Foundation, 2018 Plague Words and Phrases, http://grammar.ccc.commnet.edu/grammar/plague.htm Accessed 15 January 2019.

Crystal D 2004 *Making Sense of Grammar*, Pearson Longman, London.

Rao V, Chanock K and Krishnan L 2007 A visual guide to essay writing, Australian National University (ANU), Canberra, https://www.tcd.ie/disability/assets/doc/pdf/essayWritingVisualGuide.pdf Accessed 15 January 2019.

Rewhorn S 2018 Writing your successful literature review, *Journal of Geography in Higher Education*, 42, 1, 143–147.

Russell S 2001 *Grammar, Structure and Style*, (3rd Edn.), Oxford University Press, Oxford.

Truss L 2006 *Eats, shoots and leaves*, Penguin, London.

University of Kent 2018 Study guides, https://www.kent.ac.uk/learning/resources/study-guides.html Accessed 15 January 2019.

University of Melbourne 2018 Tertiary essay writing, Study Skills, https://services.unimelb.edu.au/academicskills/undergrads/top_resources Accessed 15 January 2019.

University of Sussex 2018 Critical essay writing, http://www.sussex.ac.uk/skillshub/?id=256 Accessed 15 January 2019.

Add up 2

Work out the total for the top of each pyramid. Answer in Chapter 28.

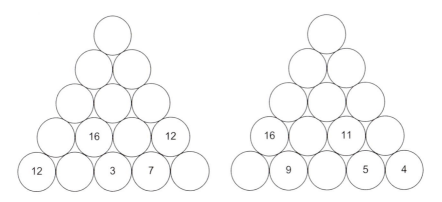

Answers: Try This 18.1 – Keywords in essay questions: some thoughts

1. *Discuss the role and responsibilities of science and scientists for society.* This essay requires a balanced four-part answer addressing roles and responsibilities for science in general and for individual scientists. Ethics, knowledge, clarity of explanation, empiricism and its limitations, oversight by communities, funders and governments are just some areas that could be included, with examples and references for each. Remember to use your discipline examples. This essay could be set for any science student; chemistry students will answer with chemistry examples, psychologists with …

2. *Communications, internal commerce and energy are the sectors that are usually identified as the most serious 'bottlenecks' in contemporary Chinese development. Explain the weaknesses in one of these sectors.* Asks for 'weaknesses' and 'one' sector only. Don't attempt to impress by covering all **three** sectors **and** strengths **and** weaknesses, giving an examiner pages of irrelevant writing to cross out. What is 'contemporary' in this context? Starting with nineteenth-century material will not help.

3. *To what extent are growth and change impeded by archaic social structures in either Latin America or a selected country in Latin America?* Key words here are 'growth', 'change' and 'impeded', so there should be little or no explanation of how social structures assist growth and change. Take examples from **either** across Latin America **or** just one country.

4. *Critically explain how Andrews (1983) derived his bedload entrainment function.* Wants a technical explanation leading to the derivation of an equation in 1983. Therefore, call on pre-1983 work and the evidence

Andrews used, and alternatives and arguments against this particular function. Any reference to post-1983 developments is beyond the scope of this essay. You might refer to subsequent developments in a closing paragraph, but very briefly.

5. *Explore the arguments that further drilling for oil should be permitted in Alaska.* This is a very open question, it asks for arguments for drilling, but do not ignore the arguments against drilling. Will you consider all of Alaska, or focus on drilling in the Alaska National Wildlife Refuge, which is topical? The perspectives of many pressure groups are important, do they have valid and evidenced positions? What are the alternatives to drilling in Alaska? This question needs you to decide on the balance of geology, resource availability, economics, politics at national and local level, ethics, landscape and wildlife impacts, …

6. *Evaluate the role of mass media reportage of environmental issues.* In 'evaluating' be complimentary as well as critical, remember the pros as in raising awareness, as well as the disadvantages. Consider all types of mass media, different types of TV reportage – the news bulletin versus the considered documentary – the print media, papers and magazines. The answer requires diverse examples. Is this a UK question or should comment be made on media activities in other countries? Incidentally, if all the lecture examples were from the UK you can use European and other examples, unless the module is country specific: 'Environmental Issues in the UK'.

Answers: Try This 18.2 – Evaluate an introduction

Some comments – do you agree? I think versions 2 and 4 are good. Versions 1 and 3 contain correct statements, but resemble a bundle of random thoughts rather than a developed argument. There is much to discuss, but version 1 says nothing about human impact on wetlands. It starts promisingly, but the last sentence is unconnected and there is nothing to indicate how the essay is structured. Version 3 could introduce a wetlands essay in any country – did you notice Canada was not mentioned? Personally, I do not like the definition of wetland in brackets in the first sentence of version 3. It is not wrong exactly, but to my mind, important enough not to be in parenthesis. The title of the essay is repeated in the third and last sentence, not ideal. Version 2 is much longer and covers considerable ground. The last sentence indicates that evidence is to be marshalled through case evidence and that there is a positive side to human impact through wetland reclamation and restoration. Version 4 is slimmer: two main points are made in the first two sentences, and then the essay structure is flagged in the last two. Versions 2 and 4 benefit from the embedded references.

Answers: Try This 18.3 – Good concluding paragraphs

Version 1 The tone here is upbeat to the point of tabloidese: 'mega' is not a good adjective and 'explosion' is OTT. The second sentence makes no sense in relation to the first, and ascribes some anthropomorphic attributes to non-sentient wetlands. The fourth sentence does not follow from the third, although using the phrase, 'This has meant', implies a logical link.

Version 2 A single-sentence paragraph is usually a bad idea. The sentiments are right but the language is very relaxed!

Version 3 Best of these!

Version 4 This version starts by sounding more like an introduction and then tails off. Individually the statements are all true, but they don't coalesce.

Version 5 Starting a final paragraph with 'To conclude' wastes words. The reader can see it is the last paragraph, The 'This essay has shown' formula lacks initiative. Not wrong, but not innovative either. The first sentence makes a geographical generalisation. Did man actually have a national policy here, or is the national picture the accumulation of a great many local, independent wetland drainage activities?

Answers: Try This 18.6 – Shorten these sentences

Wordy	Improved
In many cases, the tourists were overcharged.	Many tourists were overcharged.
Microbes are an important factor in soil processes.	Microbes control decay rates in soils.
It is rarely the case that sampling is too detailed.	Sampling is rarely too detailed.
The headman was the proud possessor of much of the land in the vicinity of the village.	The headman owned land around the village OR The headman was proud to own land near the village.
Moving to another phase of the project …	The next phase …
Chi-square is a kind of statistical test.	Chi-square is a statistical test.
One of the best ways of tackling prison reform is …	To tackle prison reform …
The investigation of cross bedding at Ramsgate continues along the lines outlined.	Investigations of cross bedding at Ramsgate are continuing.
The nature of the problem …	The problem …
Temperature is increasingly important in influencing the rate of snowmelt.	Temperature controls snowmelt rates.

One prominent feature of the landscape was the narrow valleys.	Narrow valleys are prominent landscape features.
It is sort of understood that …	It is understood that …
It is difficult enough to learn about stratigraphy without time constraints adding to the pressure.	Learning about stratigraphy is time consuming.
The body of evidence is in favour of …	The evidence supports …

Answers: Try This 18.8 – Synonyms

1 understanding, 2 main, 3 discrimination, 4 extracted, 5 explains, 6 estimates, 7 experience, 8 associated, 9 Could be either, choose morphology as the more technical term.

Not the essay: reports, laboratory and field notebooks

You are only as good as the group behind you.

Everyone writes reports using their clear thinking, analysis, synthesis, written communication and persuasive writing skills. The test of a good report is:

✓ The reader can repeat the work without reference to additional sources.

✓ The reader understands the significance of the outcomes of the work in its wider context.

✓ It is short and to the point.

The guidelines for geography students at the University of Adelaide (2018) apply to every report you write:

> 'In producing a report you are normally addressing the following questions:
> 1. What did you do?
> 2. Why did you do it?
> 3. How did you do it?
> 4. What did you find out?
> 5. What do the findings mean?
>
> Report writing is different to essay writing since:
> 1. Reports usually utilise primary rather than secondary data
> 2. They are more formally structured
> 3. You may not be required to develop an argument through the report
>
> Despite the distinction that an argument is not formally required, the best reports flow from one section to the other where the sections may be seen as a series of linked sections.'

19.1 Types of report

Consultancy report

Consultants typically provide expert reports for their clients, so it's important to understand what the client will do with the information provided. Because reports are usually used as part of the evidence base for taking a decision, they need to be clear, factual and reference all the evidence. In general the elements are:

Introduction: it explains the background and the brief the report addresses, including data collection and analysis methods.

Analysis of issues: the client is interested in the specific topics that underpin the recommendations, which determines the shape of the report. Use subheadings to clarify each issue. In each of these subsections put a brief description, your analysis of the position, possible solutions and alternatives, and recommendations.

Recommendations: collect all the recommendations together as a prioritised list. Make it easy for the reader/client to use, for example to make a decision.

Conclusions: this section should be a short summary of the issues and findings.

Appendices: detailed additional evidence, references, graphs, data, photographs …, depending on the brief.

Focus group report

Focus groups, usually involving 4–10 people, are used to collect the ideas and opinions from representative participants, usually a small sample of a population. The session is usually recorded or notes taken. Following analysis, participants' comments are selected as evidence or examples for the report. Ideally each section of the report will have two or more participants' statements. Include statements from as many different people as possible. Take care to point out that while the report properly represents the views of the group, they are a sub-sample of the population. The decision-making should take into account that results are guides rather than definitive (see Hopkins 2007). A typical report format involves: background (reasons for holding the group), optional executive summary (see Chapter 20), the main findings, and recommendations.

Project proposal or scoping report

Proposals are generally written for a manager or funder to assess the value and costs of carrying out a project. Where the decision has been taken to … (build a bridge, landscape a quarry, open a drop-in centre, research customer loyalty ...) the proposal will focus on delivering the project. If the decision has still to be made, then the first section will need to have more emphasis on why the project should be undertaken. Scoping a project involves understanding the needs of everyone who will be involved. These stakeholders need careful identification. In many environmental science and human geography projects capturing the needs of different community groups is essential but challenging.

Project proposals should convince the supervisor that the project is viable, all the risks have been assessed and can be managed, the research methods and alternatives considered, different analysis approaches have been considered, laboratory and field equipment are available, and the timescales for analysis and delivering the final report are realistic. Where people are involved, ethical clearance is sought and hazard assessments are completed for fieldwork and laboratory analysis according to university protocols.

The elements of a proposal report will vary depending on the scale of the project, but in general:

Proposal: the evidence which underpins the decision to undertake a piece of work. The project aims and objectives are clarified.

Scope and plan: the timescale, sequence of activities, risks and difficulties, people and resources required for the whole project, are identified.

Implementation stage: who is doing what and when.

Completion stage: explains the work involved at the end, the product, closing report, evaluation work, longer-term work required.

Progress report

Progress reports provide factual information on the state of a project, or on the progress of an individual or group of people. For projects taking a number of years, a progress report updates those involved and the clients on activities since the last report, and is explicit about the timeline established at the start, as being on time, ahead or delayed.

An individual may provide an employer or tutor with a reflective update on their recent activity and progress. The tone should be factual and reflective (Chapter 2), referring to previous reports, work targets and possibly your longer-term ambitions. Include: what you intended to do in the past (2, 3, 6, ...) months, what actually happened, what went right, what you might do differently next time, and your plans for the next period. Short is good.

Write your report for your audience.

19.2 Computer and laboratory practical reports

> I submitted a virtual
> report last week.

Some practicals are truly investigative, where data are analysed to find 'an answer'. You might use census data to forecast migration patterns for the next 20 years, or water chemistry analyses to explore pollution issues. The emphasis is on achieving and evaluating the answer. This type of practical is possible because you did 'developmental' practicals. Developmental practicals help you understand the research process, by evaluating where and how each

> Oh, OK, thanks, I'll
> give it a virtual mark.

step works. In this kind of investigation the report places less emphasis on the final result than on the steps involved, making statements about assumptions, validity of data, techniques, sources of error and bias, alternative approaches used and discarded, or evaluated and rejected. There may be discussion of the choice of statistical test, a forecasting model will be taken apart and its assumptions and processes questioned, alternative field approaches such as questionnaires versus interviews discussed. This kind of practical calls on your analytical thinking as well as your problem-solving skills. Depending on the type of practical investigation, your report will change its focus and emphasis. Many dissertations include elements of developmental practicals, questioning the processes and procedures used.

Report format

The normal expectation is that the writing is brief and direct, and follows the general format outlined in Figure 19.1.

Chapters/Sections	Contents
Title	Short and precise title.
Author details	Author's name, student number and contact address (email).
Abstract	Short, precise summary.
Introduction	Background. Why you are doing this.
Methodology	Computer, field and laboratory procedures, sampling methods, site details.
Results	A summary of the findings using tables, graphs and figures.
Discussion and evaluation	Interpret the results using statistics and modelling as needed. Consider the accuracy and representativeness of the results and their interpretation.
Conclusion	A brief summary of the outcomes. Describe what could be done to continue the research.
References	Vital, remember to include methodological references too.

Figure 19.1 Outline for a report

Abstract or summary

You might write this in the style of an executive summary (see Chapter 20). Explain the broader context (where this work sits within palaeogeomorphology/youth studies/spatial analysis), the hypotheses tested, methodology, main findings and interpretation in brief sentences.

Introduction

Describe why you carried out this piece of work and give a brief indication of where it fits with the rest of GEES-related information and data. The hypothesis being tested should be stated precisely. Refer to other experiments and research on this topic, which may involve a mini (1–5 paragraphs) literature review.

> Please state the hypothesis clearly

Methodology

The technical details describing the research approach, computer programmes, field or laboratory procedures.

Site details: A short site description and reasons for choice of site will set the scene. Use maps, grid references, sketches and photographs so the reader can locate and picture the location.

Sampling methods: Outline the sampling and analytical scheme. Include maps, flow charts and equipment diagrams. This is the point to discuss potential errors, calibration issues, and the influence that the selected sampling procedure might have on the results.

> Be clear about:
>
> **Accuracy**: how close a measured value is to the actual (true) value.
>
> **Precision**: how close measured values are to each other.
>
> **Bias**: systematic (built in) error which makes all measurements wrong by a certain amount.

Standard and non-standard procedures: Many computer programmes, statistical and laboratory procedures are standard. They must be cited, but do not need to be explained in detail. Statements like 'Questionnaire responses were coded and entered for analysis using SPSS', or 'The standard procedure for water hardness by titration was followed (Geography Laboratory Manual Test No 65)', or 'The programme was written in Grass and imported to ARCINFO for testing', will do fine. You do not need to detail SPSS procedures or how you tested the water. BUT if you create a new method, adapt a standard method or adopt a non-standard approach, it must be explained in enough detail for the next researcher to evaluate and copy your technique. A full technical or laboratory protocol might be put in an appendix. Explain why it was necessary to invent a new method; what was the problem with previous approaches?

Please note: Science reports often correctly include laboratory procedures, chemical lists, apparatus details and sample size information, BUT omit details of the subsequent qualitative, graphical or statistical analysis … which loses marks!

Results

Summarise the results in tables, graphs, flow diagrams and maps to show the main findings.
Label and order the diagrams so the sequence

> Support all arguments
> with evidence

flows match the logic of the discussion and evaluation sections. Consider putting the raw data in an appendix, or submit on disk; don't clutter this section with it.

Discussion and evaluation

This section will cross-reference to the results section, with the emphasis on explaining relationships, lack of relationships and patterns in the results. Consider how the results fit with previous researchers' experience. Refer back to the introductory and literature sections to place these results in their wider sub-discipline context.

Caution: There is usually more than one explanation for most things. All measurements are prone to error and inaccuracy; consider what might be a problematic, talk to people to develop some ideas. No self-respecting examiner will ever believe that you have a perfectly accurate dataset. Unless you are absolutely certain use cautious phrases like, 'It would therefore appear, on the basis of this limited data, that …', or, 'Initially the conclusion that … appears to be justified, but further investigations are required to add evidence to this preliminary statement'.

Conclusion

Resist at all costs the temptation to explain all the Wonders of the Known World from two measurements of pH in a mixture of mashed leaf, a $1cm^3$ rock sample, or a simulation of the 'Changes in population of the village of Pigsmightflybury 2022–25'. Ensure your conclusions are justified by the data, not just what you hoped to find. This is the place to suggest what research should be done to continue the investigations.

References

A report is a piece of academic work. Reference your sources and quotations fully (see Chapter 14).

Appendices

Some information is too detailed or tangential for the main sections, and distracts the reader from the exciting storyline, but it may be included in appendices. Appendices are the home for questionnaires, details of laboratory protocols, example calculations, programmes and additional data, maps and graphs. Appendices are not 'dumping grounds' for all raw data and field notes. Limit appendices as far as possible. For most laboratory reports they are totally unnecessary.

19.3 Laboratory notebooks

Many GEES practicals are assessed through laboratory reports or notebooks, which are written during the practical and submitted as the laboratory class ends. Where online lab notebooks are used this paragraph may be redundant (!) but the advice on structure and content applies to digital data records too. In-class recording encourages the adoption of professional laboratory practice, noting experimental problems as they happen and making interpretations at the time. Getting good marks probably, almost certainly, indubitably, means reading handouts and doing the recommended pre-practical reading, so that you confidently interpret the results as they arrive. A practical class is short, making this approach seem tough, but it develops your skill in describing and analysing results at speed, describing what you actually see rather than a fuzzy vision partially recalled three weeks later. If laboratory notebooks are new to you, here are some basic guidelines.

A hardcopy notebook

Your module information may specify the notebook style (possibly blank and lined pages, carbon, or graph paper). If not, select a large (A4) hard-backed notebook with a mix of lined and blank pages. Number the pages. It's an important practice to cross out any entry you are unhappy with; do not tear out a page. Lab notebooks are written in pen, and dated, so that it is clear what are your first and later records. Structure the initial pages:

Page 1	Name, email address, student number, module name and number, project name.
Pages 2–3	They will contain your Table of Contents, add to it at each session.
Pages 4–5	Leave blank at the start. It gives you space to add a summary, introduction or key as needed.
Page 6	Start writing here. Aim to record experiments, starting each one on a new page and cross-reference to the Contents page.

Recording experiments

Primarily laboratory note-keeping involves being organised and careful. Unless given specific advice follow this format consistently for each entry: title; introduction and aims; methods and procedures; data obtained; issues and problems; summary of outcomes. Have enough detail so that another student or your tutor can follow exactly what you did, knows the dates and times you worked, and there is a record of problems and issues. For some modules your work will be supervised and your laboratory record signed as a correct record of activities by your tutor, laboratory technicians or fellow students.

It is good practice to record more detail than you will need for your report. You are keeping records so that you can be very precise about instrumentation, time of day, other people helping …; if problems arise you can refer back to the notebook for accurate details.

Before you leave the lab take a few minutes to read through your laboratory notes, check that you have consistent records, that each experiment has been introduced and the outcomes summarised. It is very easy to forget to put dates on entries, to include the main results but not the equipment list, experimental set-up, calibration results …, and … because everybody forgets things.

Laboratory notebooks are assessed because they help you to develop as a professional scientist or social scientist.

19.4 Field notebooks – geologists especially

All GEES students make notes in the field. Students on geology courses will get special training in their use because professional geologists use their field notebooks (digital or hardcopy) as their professional record of activities. These are very important documents which must be accurate; they are the evidence if a client disputes field results. Getting practice as an undergraduate is very useful for professional work in later life. Creating good notebooks gives you practice in summarising information. This is particularly relevant for those seeking careers in earth sciences, forestry, planning, garden and landscape design, ecology and the construction industry; and useful in any activity requiring record keeping.

On some geography and environmental sciences field classes your field notebook will be assessed; on geology field trips it will almost always be assessed. Remember it's the notes made in the field that will be assessed. Modules in earth sciences in particular give you the opportunity to improve your field sketching and summarising skills (Chapters 26 and 27). Expect a tutor to ask to look at your field notebook during a field class, to show you how to improve your sketching style and note-making. Really good field teachers will complete their own field notebooks during a field class to show you how a professional geologist, environmental scientist or geographer makes records at a field site. Take advantage of any opportunity to see a professional's field notebook. (You will discover that most people are really bad at sketching, but are making an enormous number of notes on each diagram.)

Your main ambition is to make records, draw sketches, make notes of features, people and activities, and to slot in photographs so that you or somebody else visiting the site at a later date understands what you have done, and can build on your results.

Picking a notebook

Select a notebook that is hard-backed for writing notes in any conditions. Make sure it fits in your jacket pocket, and has a waterproof cover because it will rain. Use pencils which will write on wet days and are good for sketching, as well as pens and erasers. Ink runs in the rain, pencils are vital. Find a large clear plastic bag, big enough so you can get your hand inside to write in your notebook when it is really wet.

Digital recording

You need extra power packs, waterproof covers and don't rely on Wi-Fi being available.

Contents

Some modules have very specific rules for field notebook layouts and completion, otherwise number all the pages and add:

Page 1	Your name, contact email and address (a surprising number of field notebooks get lost (left in the pub)), module number, field location and dates.
Pages 2–3	Contents – leave blank initially, add details later.
Pages 4–5	Leave blank initially to add … later. Usually an introduction or summary.

For each visit/location you need: location information including map or GPS reference, date, names of people for group work, weather conditions as standard items. You may want to write a short paragraph outlining your aims for the session/visit, the hypothesis to be tested … Then you need as much information as possible so that you can interpret it at another time.

Field sketches get better with practice (see Williams 2018 for examples). The trick is to make sure that every picture is very well labelled. Practise landscape sketches, cross-sections and 3D sections as appropriate to your site (Chapter 27). In complicated landscapes, it is better to have five sketches each showing a particular feature with its components identified, than one very confused sketch. Before you leave a site, check that you have the map or GPS reference for your location, a compass direction arrow, an idea of the scale, slope angle if appropriate, and details of each feature which is important. Provide a key to explain abbreviations in either the front or back of your field notebook.

As with all skills, keeping good field notebooks develops with practice. People's ability to create sections and cross-sections really improves after a few visits to quarries or cliffs where the geological structure is obvious. Field sketching can be practised every time you see an urban or rural landscape.

19.5 Where the marks go!

Most students hope most marks are awarded for all the time spent sweating over a steamy laptop, talking to billions of shoppers persuading them to complete questionnaires, trying to log into remote computers, digging trenches across moraines and mixing sewage sludge in acid. TRAGICALLY, most academics see these efforts as worthy of very minor reward. Practical sessions develop your skills and experience in handling data, samples or spades, and are worthy of maybe 50 per cent of the marks at the most generous. Your tutor produced handouts, briefings and explained what to do. What s/he really wants to know is:

- Have you understood what the results mean?
- Do you know why a geographer or geophysicist would want to be doing this in the first place?
- Do you understand the flaws in the method and the accuracy of the results?
- Can you explain what an intelligent environmental scientist would research next?

Know the deadlines!

Most marks are assigned to the discussion, evaluation and conclusions section of a report. Leave time to write these up and redraft them so the writing is good and *add the references*. Many reports lose marks because they stop at the results stage, missing the interpretation.

> Collaborate – share drafts to get feedback from friends and study buddies.

This next sentence completed a student report: '*It can therefore be concluded that a fuller investigation is required, increasing both the sample size and its spatial extent*'. This conclusion was on the right lines and showed the author appreciated that sampling was an issue BUT it is possible for the author to add value with a conclusion that is more information rich. Something like: 'The restricted data set has limited the inferences that may be drawn. Follow-up research should incorporate a broader cross-section of respondents. If finance were available, extending the research to contrast small towns (Bridgwater, Colchester, Ledbury and Thirsk) with major industrial and commercial cities (Bristol, Glasgow, Leeds and Manchester) would increase confidence in the conclusions'. This is longer but no more difficult to write. The original author had discussed future experimental work informally, earlier in the project, but said '*I was shy of adding these sorts of ideas to the report*'. Shy can be nice but doesn't get marks. Forget 'shy' when reporting.

Use your departmental practical assessment sheets as a guide to self-assess and revise your reports, or use Figure 19.2. A 60/30/10 mark split is indicated here; what counts in your department?

Names ....

Laboratory Report Title ....

	1	2:1	2:2	3	Fail	
Structure and Argument (60%)						
Logical presentation Topics covered in depth Links to (broader discipline) clearly made Clear succinct writing						Discontinuous report Superficial report No links made to (broader discipline) Rambling, repetitious report
Technical Content (30%)						
Experiment appropriately described Graphs, tables and maps fully and correctly labelled Sources fully acknowledged Correct, consistent use of units						Experiment poorly described Incorrect/no labelling of graphs, tables or maps No references Inconsistent use of units
Presentation (10%)						
Good graphics High standard of English						Poor use of graphics Limited fluency in writing
Feedback						

Figure 19.2 Self-assess a laboratory report

Drawing conclusions

It can be very tempting to draw firm conclusions from data. It's important at the end of a project to consider (*think through*) the evidence accumulated and its wider value. Any student project explores a limited area because there is neither the time nor the resources to work on something for a number of years. Be circumspect. Check the adjectives used in your discussion and conclusion. If you're using words from the left-hand column, might the conclusion be better expressed using phrases from the right-hand column?

Unarguably we can see	This limited evidence suggests
Certainly	Possibly/maybe
Unquestionably	Perhaps
Absolutely confirmed	Lends some evidence to
Positively conclude	Tentatively conclude
Definitely indicates	Lends cautious support to

19.6 References and resources

Hay I 2012 Writing a report, in Hay I *Communicating in geography and the environmental sciences*, (4th Edn.), Oxford University Press, Melbourne, 30–56.

Hopkins C 2007 Thinking critically and creatively about focus groups, *Area*, 39, 4, 528–535, https://doi.org/10.1111/j.1475-4762.2007.00766.x Accessed 16 January 2019.

Lewis S and Mills C 2007 Field notes: a student's guide, *Journal of Geography in Higher Education*, 9, 1, 111–114.

Science Buddies 2018 Science and Engineering Project Laboratory Notebooks, https://www.sciencebuddies.org/science-fair-projects/science-fair/laboratory-notebooks-stem#usingalabnotebook Accessed 16 January 2019.

University of Adelaide 2018 Geography report writing guidelines, https://www.adelaide.edu.au/writingcentre/sites/default/files/docs/learningguide-geographyreportwriting.pdf Accessed 16 January 2019.

Williams, P 2018 Fieldwork notebook tips, http://www.tjpeterwilliams.co.uk/fws/tips.htm Accessed 16 January 2019.

Be aware of oxymorons

Can you add to this collection? See Chapter 28 for a few more.

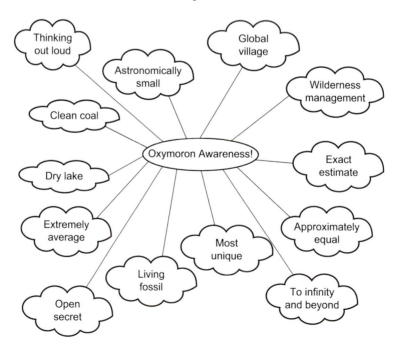

20

Abstracts and executive summaries

I got the abstract sorted yesterday ... so that just leaves the dissertation.

Most journal articles start with an abstract that summarises the contents. Executive summaries are normally found at the start of reports and plans, particularly business documents. An executive summary (ES) aims to describe the essential points within a document, usually in one or two pages. Depending on the context, the style may be more dynamic and less formal than an abstract.

20.1 Abstracts

Look at some well-written abstracts before writing one. Most journal articles start with the abstract – see *Energy and Environment, Transactions of the Institute of British Geographers, Hydrology and Earth System Sciences*, or *Journal of Geophysical Research* for examples.

An abstract should be a short, accurate, objective summary including key findings; there is no room for interpretation or criticism. Abstracts should do the following:

☺ let the reader select relevant documents for a particular research problem;

☺ substitute, in a limited way, for the original document when accessing the original is impossible.

Start with **Try This 20.1**.

TRY THIS 20.1 – Abstracts

Next time you read a journal article, read the paper first and make notes without looking at the abstract. Then compare your notes with the abstract. Are there significant differences between them? How can you use an abstract as a summary?

Remember: reading abstracts is not a substitute for reading the whole article.

When asked to prepare an abstract of some text as a tutorial exercise, aim for 80–150 words:

✓ give the citation in full;

✓ lay out the principal arguments following the order in the text;

✓ emphasise the important points, highlight new information, omit well-known material;

✓ be as brief, but as complete, as possible;

✓ avoid repetition and ambiguity, use short sentences and technical terms;

✓ include the author's principal interpretations and conclusions.

✗ do not add your own commentary. This is not a 'critical' essay.

The first draft will probably be too long and need editing.

20.2 Executive summaries

An effective executive summary (ES) is a shortened version of a document aimed at a more general audience. They are often found in reports (Chapter 19). A business may produce an ES for the general public, local authority or promotional purposes. ESs are generally less literary than abstracts, with bullet points or numbered sections. The general rule on length is one side only, on the basis that really, really, really busy people will not read more. For promotional purposes an upbeat, clear style with lots of impact is advantageous. The iWrite (2014) link shows three student ES examples with markers' comments – well worth a look.

An ES for a student report or project is normally one side or 350 words. Eliminate all extraneous material. Do not include (severely limit) examples, analogous material, witticisms, pictures, diagrams, figures, appendices, or be repetitious or repetitious or repetitious. An ES should:

✓ be brief and direct;

✓ include all the main issues;

✓ indicate impacts, pros and cons;

✓ place stress on results and conclusions;

✓ include recommendations with relevant costs and timescales.

Look at the discussion and conclusion sections for the main points. Find examples of ESs through **Try This 20.2**. Geography students, set the task of generating an ES, thought it looked remarkably like a good outline plan, albeit with the emphasis on results and recommendations.

TRY THIS 20.2 – Executive summaries

Search the web for ES examples using 'executive summary …' and any topic you choose. Adding 2019 will prioritise examples from that year, or start with these (all sites accessed January 2019):

Deloitte 2018 The top 10 issues shaping mining in the year ahead, https://www2. deloitte.com/content/dam/Deloitte/global/Documents/Energy-and-Resources/gx-TTT-executive-summary-2018.PDF

UNEP 2017 The Eighth UN Environment Emissions Gap Report, http://wedocs.unep.org/ handle/20.500.11822/22101

World Economic Forum 2018 The Global Risks Report p 6, http://www3.weforum.org/ docs/WEF_GRR18_Report.pdf

Yale University 2018 Environmental Performance Index, https://epi.envirocenter.yale. edu/2018-epi-report/executive-summary

---

Using abbreviations for longer words or phrases repeated regularly can speed up writing. How do you view the use of 'ES' rather than 'executive summary' in this chapter? Should this style be adopted? Note that when the phrase appears in the first paragraph, it is followed by '(ES)' to indicate that this abbreviation will be used thereafter.

20.3 References and resources

iWrite 2014 Executive summary examples, University of Sydney, http://iwrite. sydney.edu.au/ENGINEERING/Reports/Group-Project-Report/Executive-Summary-examples.html Accessed 16 January 2019.

SkillsYouNeed 2018 How to Write an Executive Summary, https://www. skillsyouneed.com/write/executive-summary.html Accessed 16 January 2019.

UniLearning 2000 Examples of good and poor executive summaries or abstracts, https://unilearning.uow.edu.au/report/3bii1.html Accessed 16 January 2019.

Words in geo-words 3

Set eight minutes on the timer (cooker, phone …) and see how many words you can make from

FIELDWORK

☺ Thirty is a very good score. See Chapter 28 for answers.

21

Reviewing papers and books

Authors love criticism, just so long as it is unadulterated praise.

The process of critically reviewing papers or a book helps develop understanding of the broader context of a topic, and provides practice in expressing your thoughts and opinions. Completing a review assignment effectively involves making the time to understand both content and the wider picture. Ideally, allow a week to read the paper or book, a couple of weeks to read other material, think and then write. Use regular writing sessions to prepare – Chapter 5.9.

> **Reviewing:** making a critical appraisal of information (book, paper, chapter, web resource).
>
> Skills involved include: analysis, appraisal, assessment, critique.
>
> You comment, criticise, discuss, evaluate, inspect, investigate, judge, rate, reflect on, weigh up.

These tutorial or module assignments provide practice for larger projects and dissertations. Critically evaluating the relevant literature, and setting your research in its broader context is an essential, albeit brief, element of large projects.

A good review provides an objective summary of contents, scope, approach and importance of the book. Aim to evaluate its quality, limitations and applicability in its wider context. Remember reviews are personal, so make notes of your thoughts as well as the subject material. For examples see the resources section (21.3).

21.1 Reviewing papers

Reviewing a selection of academic papers allows you to evaluate ideas, issues and results and put together a new thesis or argument which presents a particular point of view. Plan time to research and read, think creatively, construct an argument and write your review.

Read some recent reviews first!

Reviews of papers usually focus on the development of ideas, knowledge and applications. You might review papers, possibly from different disciplines, to

A review **can't** be completed in one draft, the night before a tutorial

evaluate a methodological or experimental approach. With either challenge, your initial reading should focus on both the development of ideas and applications, and the methodology, and then decide the focus and balance of your review.

Your approach could involve:

- Making notes when reading, starting with a brief summary of the contents (the abstract may not be a full summary).
- Highlighting references to previous work cited by the author to read for background; and reading them!
- Searching online for more recent studies; select some to read, and read them!
- Thinking through the context. Journal articles have word limits, so authors are selective.
- Ask whether the conclusions fit with the evidence?
- Consider where this paper fits into the development of ideas, knowledge, application and methodologies? What other evidence might have been used?
- Draft the review; think; re-check your notes; edit and revise; include references.

Reviewing involves giving yourself enough time to absorb content and let your brain make connections to different ideas and sources so that your links, compliments and criticisms are compelling.

21.2 Book reviews

Reviewing a book allows a broad theme to be considered. Think constructively around what is included and what is omitted. Be impartial. Counterbalance any weakness or imbalance with praise for the strong sections. Authors have to make choices and usually justify them, but it helps the reader to know both what is and is not included.

Use quotes appropriately and reference them at the end. Long quotations are not a good idea unless they really illustrate a point. Including examples and evidence from alternative texts can add richness and balance. If it is well written, and good value, please say so, '*this text is clearly written, with case studies and examples illustrating key points undergraduates will find useful*', or '*… the writing is turgid and difficult to understand*'.

Your approach could involve:

- Skip-reading the book to understand its scope.
- Identifying the author's aims and intended audience; does the book deliver effectively to the intended audience?
- Making a brief summary of the contents.
- Spotting quotations or references to illustrate your review.

📖 Locating references to papers and books to read to put this work into context; then find and read them! Check the library for more recent books and resources.

📖 Your thoughts on any significant areas omitted.

📖 Deciding the theme for the review; draft the review in outline; check your notes for quotations and ensure nothing vital is missed.

📖 Complete your review, edit and revise; check all references are included.

21.3 Resources

Environmental Economics 2018 Book Review blogs, http://www.env-econ.net/book_reviews/ Accessed 15 January 2019.

Resurgence and Ecologist 2018 Book Reviews, https://www.resurgence.org/magazine/book-reviews.html Accessed January 2019.

Scottish Journal of Geology 2019 Book Reviews [in most volumes], http://sjg.lyellcollection.org/content/current Accessed 15 January 2019.

Society for Urban Ecology 2018 Book Reviews, https://societyforhumanecology.org/book-reviews/ Accessed 15 January 2019.

Urban Geography Research Group 2017 Book Reviews, http://urban-geography.org.uk/book-reviews Accessed 15 January 2019.

Geogram pairing 2

Link each word in the left-hand column to a word in the right-hand column which is its anagram. Answers in Chapter 28.

boatel	albedo
clast	course
doable	sleet
inhale	talcs
nacre	master
source	haline
stream	filter
tablet	oblate
teles	battle
trifle	crane

22

Presentations

Tension's what you pay when someone's talking.

GEES degrees are littered with speaking opportunities. This is a really good thing! Somewhere on your CV you can write, *'During my university career I have made 25 presentations to audiences ranging from 5 to 165 using PowerPoint with embedded video and podcasts'*. This impresses employers. Many GEES graduates forget to tell employers that they have had these opportunities. Presenting helps you to manage your nerves and allows you to become familiar with the question-and-answer session that follows. Grab all opportunities to practise your presentation skills. They all count for CVs, presentations in tutorials, seminars and mini lectures, on podcasts and video.

This chapter has useful advice, but presentations are essentially a visual performance that needs practice. Think about your lecturers. What do you want to avoid doing? What would be good to do? Check out YouTube: *How to give … funny/ knockout/effective … presentations*, and TED talks.

22.1 Types of presentation

The design and emphasis of your talk changes depending on your aim and audience. Clarifying their expertise and reasons for listening before you start is exceptionally helpful. Are you:

✓ **Delivering information:** Are you explaining or do you also have to persuade? Include facts, examples, case studies and data but leave time for people to absorb the new ideas. Check people understand and expect questions at the end. Be logical and back up your ideas with evidence (*references*).

✓ **Demonstrating or teaching a process or skill:** These sessions are most effective with active participation, where the audience is involved, trying out and experimenting. Be encouraging. Thank your audience for getting involved.

✓ **Giving a progress update:** Important as part of project work. Refer to the original plan and timeline, the expected and actual progress. Outline the project's next steps.

✓ **Selling a product or service:** You need to really understand the product or service first. How would your audience use it? What are the benefits for them? A good sales pitch concentrates on why the product should be used, cost and delivery issues.

✓ **Presenting at an interview:** This is a normal part of a recruitment process. Looking smart and business-like is important. Prepare properly to be super confident. See your university Careers Service website, University of Kent (2018) and Young (2018) for advice.

✓ **Webcast, Skype, web conferencing:** These require some extra preparation because you are detached from your audience. A practice run with a friend face-to-face is important. See Chapter 9.3 for telephone skills advice. Keeping to the point is very important on the phone because you cannot see how the other person is responding to you.

22.2 Getting started

You need a well-argued and supported (*add references*) message to enrapture (ideally) the audience. Other chapters explain how to get the information together; this one offers practical presentation skills. The four most important tips are:

1. Suit the style and technical content of your talk to the skills and interests of the audience. Make the content accessible, so that the audience wants to listen and ask questions.
2. Buzan (2018) shows that people are most likely to remember:
   (a) Items from the start of a learning activity.
   (b) Items from the end of a learning activity.
   (c) Items associated with things or patterns already learned.
   (d) Items that are emphasised or highlighted as unique or unusual.
   (e) Items of personal interest to the learner.
   Tailor your presentation accordingly. Help the audience to think and understand.
3. Get the message straight in your head, organised and ready to flow. Lack of confidence in the content → insecure speaker → inattentive audience → bad presentation = low marks.
4. Remember, the audience only gets one chance to hear you. Keep a short presentation short, essential points only. Make clear links between points and add brief, but strong, supporting evidence (*references* are good). Look cheerful because you are really happy to be there (acting may be necessary).

You must have a plan, and stick to it. Basically it is down to PBIGBEM (Put Brain In Gear Before Engaging Mouth).

TOP TIPS

On style:

→ **Can I read it?** Reading a script will bore both you and the audience. Really useful tutors (you will hate him/her at the time) will remove detailed notes, letting you 'tell the story in your words'. You are allowed bullet points on cards to remind you of the main points, but that is all. Illustrate your talk with maps, equations and diagrams, but remember that reading PowerPoint slides aloud is another cop-out.

> All slides must be visible from the back of the room.

→ **Language.** A formal presentation requires formal speaking. Minimise colloquial language, acronyms and paraphrasing, and limit verbal mannerisms like the excessive use of hmm, umm, err, neat, cool, wicked and I mean. Help the audience by putting new words or acronyms on a handout or slide.

→ **Look into their eyes!** Look at the audience and smile at them. If they feel you are enthusiastic and involved with the material, they will be more involved and interested.

→ **How fast?** Not too fast, slower than normal speech, because people taking notes need time to absorb your ideas, to get them into their brains and onto paper. You know it, but it is new for your audience. Watch the audience to check if they are lost (too fast) or falling asleep (too slow).

→ **Stand or sit?** Sitting encourages the audience, it feels less formal and encourages audience participation.

→ **How loud?** So that the audience can hear, but do not feel you are shouting. Ask someone you trust at the back to wave if volume change is needed – getting it right takes practice. Record yourself on your phone or computer. When you have finished laughing at the result, decide whether you speak at the same pace and pitch all the time. Getting excited about the material, showing enthusiasm helps to keep your audience attentive and involved.

→ **Repeating points?** Emphasise important points by repetition, or simultaneously putting them on a slide or flip chart.

→ **Handouts?** Yes, ideally one page with the main points, sometimes a complicated diagram and a few references. Depending on circumstances you may circulate this in advance (using the flipped classroom approach).

22.3 Crutches (slides, flip charts, video …)

Audiences need to understand your message, so the visual elements should support, not take over. Visual assistance can include some or all of:

☺ A title slide with your name and email address, so the audience knows who you
are and where to find you!

☺ A brief outline of the talk: bullet points are ideal, but adding pictures is fun –
just make sure the message is clear. Is Figure 22.1 clear enough or OTT? (For a
ten-minute presentation it is too ambitious. There are too many pictures. The
punctuation is muddled. Capitals must be used consistently.)

☺ A map – so the audience knows where you are.

☺ Graphs, cross-sections, and pictures. Colours on complex diagrams help to
disentangle the story.

☺ Videos – including other people's work is fine if it makes your point effectively,
but you probably only need a short section. Look at a couple of videos (try Sea
Change Project, 2018 or UN, 2018 sustainability goals) and consider what
might be selected to make specific points.

☺ A final summary slide. Possibly the second outline slide shown again, or a list
of the points you want the audience to remember.

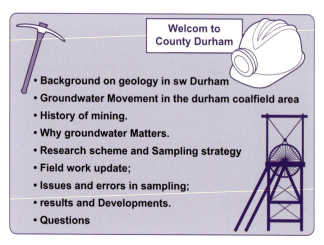

Figure 22.1 Slide 2 – Setting the scene for a ten-minute presentation; spot the typos and format errors that
need correcting.

PowerPoint slides

Preparing effective slides takes practice. What does not work: small writing, too
many colours, words from edge to edge – leave a margin. Misuse of, or inconsistent
Use OF capitals Doesn't help Either.

Good practice includes:
- ✔ Slides prepared carefully in advance.
- ✔ SPELL CHECKING EVERYTHING.
- ✔ Minimum font size 24. Keep writing **VERY BIG** and messages short!
- ✔ Adding video, cartoons, clipart items or photographs makes the message more memorable, they really help with very technical or dull messages. Make sure visual items are totally relevant.
- ✔ Checking that hot links to websites and videos are working before the talk, and that sound on the video is OK.
- ✔ Using diagrams and maps from books and papers; scan them BUT you must cite their source (Chapter 14).
- ✔ Some audience members are colour blind. Avoid green and red together, black with blue, and black on red. Yellow and orange do not show up well in large lecture theatres, and orange and brown are not easy to distinguish from 30 rows back.
- ✔ When you have lots of information and time is short, provide a handout with the detailed material and use the talk to summarise the main points.
- ✔ Reference every source, this especially includes photographs (Figure 22.2).

Before giving a talk, investigate the room, computer, projection system and microphones. Find the on/off switches, the plugs, light switches, and passwords in case the computer crashes and needs rebooting. 'Sod's law of presentations' says the previous speaker will breeze in, deliver a brilliant talk and turn everything off as s/he finishes leaving you to reset the computer while the audience watches you! KEEP COOL and DON'T PANIC.

Run the slides through in advance because sometimes the settings need changing to ensure slides show OK, can be read easily and that video and internet links are working. In brightly lit rooms, background colours may need altering at the last minute.

Figure 1 Athabasca glacier, British Columbia

Figure 1 Athabasca glacier, British Columbia. (Personal photo, Kneale 2009)

Figure 22.2 Acceptable and unacceptable PowerPoint slides. Always acknowledge sources.

Flip charts

Flip charts are an excellent way to capture thoughts in teamwork and discussion sessions. They can be pre-prepared with main points and diagrams. **WRITE BIG** and make sure your pens are full of ink. Left- and right-handed persons need the chart in slightly different positions. Check beforehand what the audience can see and adjust your position accordingly.

(BTW, if spelling is a problem, flip charts are a BAD IDEA. The audience will remember your spelling error rather than your message.)

22.4 What to avoid

☹ **Getting uptight.** All speakers are nervous. THIS IS NORMAL. Take deep breaths and relax. Being well prepared leaves time to walk slowly to the room and have a coffee beforehand. Get there early enough to find the loos, lights and seats. Ensure that the IT, flip chart, handouts, notes and a pen are in the right place for you and the audience. Take a lot of deep breaths. YOU WILL BE FINE.

☹ **Showing the audience you are nervous.** Ask your mates to tell you what you do when speaking. Everyone gets uptight, but fiddling with hair, pockets, clothes, keys, pencil, ears and fingers distracts the audience who watch the mannerisms and forget the talk.

☹ **Overrunning.** Reckon you can speak at 100–120 words per minute. Practise in advance with a stopwatch (on phone, tablet, cooker), reducing the time by 10 per cent.

22.5 Handling sticky situations

☺ Ignore late arrivals, unless they apologise to you, in which case, smile your thanks. Starting again for them interrupts your flow, irritates people who arrived on time, and encourages late arrivals.

☺ Time is up and you have 20 more things to say. This generally happens if you haven't practised aloud. You can read much, MUCH faster than you can speak. But if time runs out either skip straight to your concluding slide OR mention the missing main points (headline style) and do a one-line conclusion.

☺ Good question, no idea of the answer. Say so. A phrase like '*That's a brilliant thought, not occurred to me at all, does anyone else know?*' will cheer up the questioner and hopefully get others talking. Grab thinking time by saying: '*I am glad this has come up …*', '*I wanted to follow up that idea but couldn't find anything in the library, can anyone help here?*' and '*Great question, no idea, how do we find out?*'

☺ The foot–mouth interface. They happen. GEES favourites include '*… improved the engineering of buildings so they stood a better chance of collapsing in an earthquake*'; '*The gravity model can be heavy going to programme*'; '*What we found was that the desert was really, really dry, quite arid in fact*'.

☺ Make notes during Q&A sessions, it's all useful feedback.

⇨ TOP TIPS

→ Self-assess your bathroom rehearsals against your department's presentation feedback form, or use Figure 22.3 to polish a performance.

→ Watch your language. Strive to use a formal, technically rich, academic style. Not: '*… results seemed a bit unlikely*', '*There were loads of rocks …*' or '*Clouds, mmm, yes, the fluffy ones*'.

→ Practise your talk on the parrot, the bathroom wall and … Speaking the words aloud will make you feel happier when going for Gold. You can pronounce unfamiliar words (equifinality, fluorapatite or *Rhytidiadelphus squarrosus*) privately at the bus stop. When in doubt ask someone, the librarian, your mum, anyone.

→ Most marks are given for content, so research, research and *add references*.

Oral Presentation					
Name .... Title of Talk					
		Great	**Middling**	**Oh No!**	
Speed	Spot on				Too fast/slow
Audibility	Clear and distinct				Indistinct
Holding attention	Engaged audience				Everyone asleep
Organisation	Organised logical structure				Scrappy and disorganised points
Relevance	All material on topic				Random and off the point
Academic support	References for factual and visual materials				No references
Slide production	High-quality clear and referenced				Unclear and unreferenced
Handling questions	Thoughtful answers				Limited ability to extend the discussion
Teamwork	Good balance of team input and questions				Unequal and unbalanced presentation

Figure 22.3 Self-assess a presentation (ideally use your department's version)

22.6 References and resources

Watch TED talks and YouTube presentations. Search for *public speaking, oral presentations, great openings, speaking skills, flipchart top tips, using PowerPoint and interview presentations.*

Buzan T 2018 *Mind Map Mastery: The complete guide to learning and using the most powerful thinking*, Watkins Publishing, London.

Powell M 2011 How to open and close presentations? https://www.youtube.com/watch?v=Yl_FJAOcFgQ Accessed 16 January 2019.

Sea Change Project 2018 Sea Change Resources, videos, http://seachangeproject.eu/ Accessed 16 January 2019.

SGSB 2017 Use Body Language to Rock Your Next Presentation, Stanford Graduate School of Business, https://www.youtube.com/watch?v=pp4YlyXjcKI Accessed 16 January 2019.

UN 2018 United Nations Sustainability Goals, https://www.un.org/sustainabledevelopment/news/videos/ Accessed 16 January 2019.

University of Kent 2018 Tips on Making Presentations, Careers and Employability Service, https://www.kent.ac.uk/careers/presentationskills.htm Accessed 16 January 2019.

University of Sussex 2016 Presentations, SussexSkillsHub, http://www.sussex.ac.uk/skillshub/?id=258 Accessed 16 January 2019.

Young, K 2018 Four tips for delivering the best interview presentation possible, Hays, https://social.hays.com/2017/04/26/four-tips-for-your-interview-presentation/ Accessed 16 January 2019.

Wordsearch 3

Just 30 countries to locate here. Answers in Chapter 28.

Angola, Algeria, Botswana, Burundi, Cameroon, Chad, Congo, Djibouti, Egypt, Eritrea, Ethiopia, Gabon, Gambia, Ghana, Kenya, Lesotho, Libya, Malawi, Morocco, Mozambique, Namibia, Nigeria, Senegal, Somalia, Sudan, Togo, Tunisia, Uganda, Zambia, Zimbabwe

```
M  A  A  A  N  R  C  U  O  T  T  T  O  A  A
O  E  N  I  I  L  T  C  P  U  C  C  I  T  N
Z  R  G  L  G  D  C  Y  N  A  C  B  H  O  Q
A  T  O  A  E  O  G  I  M  U  M  G  B  A  A
M  I  L  M  R  E  S  E  M  A  L  A  W  I  D
B  R  A  O  I  I  R  A  G  A  G  S  C  N  U
I  E  M  S  A  O  N  D  J  I  B  O  U  T  I
Q  G  O  Y  O  A  I  P  O  I  H  T  E  A  B
U  N  N  N  W  W  O  L  I  B  Y  A  A  L  U
E  E  A  S  A  W  U  G  A  N  D  A  N  G  R
K  A  T  M  S  U  D  A  N  G  M  C  A  E  U
A  O  E  A  I  B  M  A  Z  P  E  O  H  R  N
B  D  H  E  W  B  A  B  M  I  Z  N  G  I  D
T  A  G  L  A  C  I  A  D  C  O  G  E  A  I
L  E  S  O  T  H  O  A  T  O  G  O  T  S  E
```

23

Posters and display stands

Aim for short, clear, short, relevant, short, pertinent and short.

University staff and students share their research ideas and results at conferences and other events using posters. A display stand at an exhibition or conference usually involves a number of posters and additional resources, sometimes demonstrations. Traditional poster

> A poster is not an essay in very small font on one page.

formats (A0 size or a single PowerPoint slide) have limited space, which forces presenters to concentrate on the essential elements using brief, concise statements and explanations. There is no space for waffle. Whether presentations are web-based or hardcopy, presenting through posters primarily develops your ability to communicate ideas clearly and concisely. It's a super opportunity to be creative. For advice on presenting at a conference or event outside your department or university see Hill *et al.* (2017a, b).

23.1 Posters and e-posters

Staff and students' research posters are frequently displayed on corridors, office and laboratory walls. Take a critical look at them. How effectively is the 'message' communicated? Are the main points readable from 2 metres? Are you enticed into going closer and reading the detail? Do you like the colour combinations? Is there too much or too little material? Is there a good mix of images and written information?

No time to visit your department's display boards? Check out Purrington's (2017) examples of effective academic posters and the Macaulay Institute (2010) which has a set of five soils and four vegetation posters aimed at a school student audience. They have a great balance of text and pictures. BCUR (2014) has a selection of posters presented in the UK Parliament by students from many disciplines summarising their undergraduate research. Consider (think about) the style and visual impact. What elements encourage you to read further?

The fastest way to lose marks is to print an essay in 14pt font and glue it to a poster board; expect instant failure. Sound-bite-length messages are wanted. However, this is an academic exercise, so your sound bite must have a good academic argument and evidence, references and sources.

Getting the design right will help the message; you want to attract attention. See Figure 23.1 for some ideas. Posters are usually rectangular, but more jolly shapes

– a volcano, map of Iceland, a greenhouse for global warming or a county map for local planning – will attract attention. Don't let the background overwhelm the message, the background should complement and enhance not dominate. Avoid clichéd images. Amongst 50 'sustainability themed' posters there might be 45 on green boards with green, cream and brown mounts. Using a different background could make your poster stand out. Whatever you decide, ensure the maximum width and height are within size limits.

Interactive posters with pop-up effects, wheels to spin, fish that move as they migrate upstream, boreholes ejecting gases, or overlays that slide away can be very

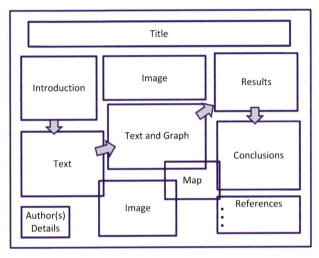

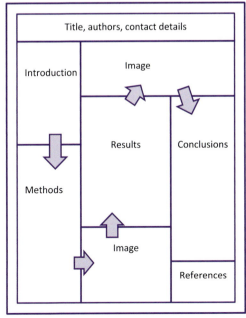

Figure 23.1 These sample layouts can be improved. What would you do?

effective. Multi-layered card is heavy. Use strong glue, reinforce cut edges and en-sure your poster can be safely attached to the wall or board.

Figure 23.2 provides a checklist for developing your poster. Design is very impor-tant. Highlight pictures and text with contrasting colours and backgrounds. Theme the colours: primary information on one colour, supplementary information on another. Be consistent in design format; it assists the reader. Let the story flow, for example by placing argument or background information to the left of an image, graph or picture, and the interpretation or result to the right. It may be effective to have a hierarchy of information with the main story in the largest type, and more detailed information in smaller types. Coloured or size-coded information consistently helps the reader to decide whether to read the main points for a general overview, or the whole in detail. Edit, edit, edit to make sure the message is clear and move sections around to find the best layout. Typical designs have space for 300–350 words, roughly one side of typed A4. Put details in a separate report or handout, and be ready to talk to everyone.

	Activity	Tasks and Issues	Deadline
1	Focus and title	Get the focus clear. Draft a title. Agree deadline dates **Read the assessment rules in the module handbook.** Specific requirements for poster size, landscape or portrait, font sizes and references are often in the 'rules'.	
2	Content	Get the academic content right. Research – discuss – research – discuss – spot the gaps – research – and then finalise the focus and title.	
3	Draft version A	Mock up a first layout. Does the title 'grab' the reader's attention? Are academic points clear? Does the design have impact? Can it can be read from 2m? Reread the assessment information or use Figure 23:3 to assess and plan improvements.	
4	Draft version B	Revise and show to other people. Ask what message it conveys. Is it what you intended? Are the reading sections in the order you wish? Are they distinguishing the major points from support material? Are the main arguments supported by clear evidence? Does the design and colour scheme help them understand the issues? Are they encouraged to read further? Redesign to eliminate anything readers find confusing.	
6	Final Version	**Recheck the guidelines and assessment rules** (or Figure 23:3). It's the last chance to change content and design to meet the criteria for a high grade. Double-check that all sources are acknowledged, and maps, graphs, diagrams and photographs have keys, scales and titles. **SPELL CHECK EVERYTHING AGAIN.** **Finalise ➜ Print ➜ Submit ➜ Celebrate**	

Figure 23.2 Key steps in producing a poster – agreeing deadlines is vital for group productions

There are costs involved in paper poster production, including card, photocopy-ing, enlarging and printing photographs. Be certain everyone is happy with the size and shape of each diagram before printing. A rough draft or mock-up, in black and white, before a final colour print, saves money. The first five to seven drafts are never right; font sizes that seem enormous on a computer screen look small on a poster board viewed from 2 metres away. Every poster produced by a group is the subject of much discussion and change before everyone is happy!

SPELL CHECK EVERYTHING EVERY TIME

E-posters can be created in Word, PowerPoint and graphics packages. They are low cost, easy to store and shared electronically. They can be created online, which is vital when groups cannot meet. The maximum size will probably be defined as part of the project, but you still need to consider the screen size of a display. Use attractive and clear typefaces and font sizes. Do you find CAPITALS EASIER TO READ or prefer Mixed Case? Different typefaces, colours and shading can highlight different types of information.

An e-poster can include links to video, animated cartoons, and short interactive computer programmes to demonstrate a model in action. There are many possibilities, but check for file size limits or software rules set as part of the assignment. Remember that all your creative innovations will get zero marks if the result can only be viewed with software assessors cannot access. Use your module guidelines or Figure 23.3 to self-assess your poster and enhance it.

Group Names............						
Poster Title............						
	1	2:1	2:2	3	Fail	
Poster Structure (20%)						
Well organised						Disorganised
Focused						No clear focus
Research and Argument (60%)						
Main points included						No grasp of the main issues
Points supported by evidence						No corroborating evidence
Good use of relevant examples						Lack of examples
Ideas clearly expressed						Muddled presentation
Design and Presentation (20%)						
Creative layout and design						Poor design
Good graphics						No/poor graphics
Good word processing						No/poor word processing
Key points readable at 2m						Key points unreadable at 2m
Feedback						

Figure 23.3 Self-assess your poster

23.2 Instant posters

When you are asked to create a poster in the next 15 minutes (or even sooner) you are getting the opportunity to practise pulling ideas together quickly, selecting material and presenting it effectively. Whatever the format (flip chart, single PowerPoint slide or web page) aim to express your ideas in as few words as possible. Let pictures, diagrams and graphs make your points.

Creating a poster to summarise the research outcomes from a fieldwork day happens on many field classes. Recognise that field days produce lots of evidence. Select and then **Edit, Edit, Edit**. Human geographers cannot include every interviewee's opinions and feedback. Choose one or two quotes that make your points, the rest will go into your report. Geology posters might include field sketches, data, graphs and, crucially, your interpretation and conclusions from your results. Make sure photographs and diagrams are properly titled, and that their relevance to your research is clear. The source of photographs must be acknowledged if taken from the internet (Chapter 14).

Studying posters created by other teams will help you to develop further ideas. Peer review and assessment of posters is usual because lecturers know that students learn much more through formative feedback from other groups than from written summative feedback after submission (Figure 23.3).

23.3 Organising an exhibition or conference stand

Organising a display for an exhibition brings together a team of people to present ideas to a broader audience, which may include academic staff, external visitors, students, friends and family. Set as an assignment, tutors are giving you the opportunity to practise developing materials as a team, creating a presentation narrative to be delivered to people visiting your stand, networking, conversational and presentation skills. Working as a team provides ideas, enthusiasm, energy and the opportunity to cover a diversity of angles, much more than somebody working alone could possibly manage. Inevitably, with a team of five or six there will be many ideas and not everything will be possible, so discussion, prioritisation, negotiation and compromise skills are involved (Hill *et al.*, 2017a, b).

An exhibition involves verbal discussion of research findings with visitors. These conversations require you to adapt your promotional 'blurb' to match visitors' interests and attitudes. This skill is different from an oral presentation for a tutorial or seminar; it develops your ability to think and adapt the message as you speak. It's helpful preparation for presenting your ideas, for example when proposing research projects or at interviews.

Take a critical look at an exhibition, for example at university Open Days and Careers events. Employers promote their organisation with clear, attractive

> *It was easier to talk to one person than a whole room, as you can adjust to their reactions and feel they are listening.*

messages. Note what you like and dislike: materials, size of photographs, fonts, colour scheme, background and handouts. What enticed you to a stand? What encouraged you to talk with exhibitors? Was it the free tea or toffees? What can you adopt for your project?

In designing the stand ensure there is space for all of the team to talk with visitors. There should be a logical arrangement to your materials so that a visitor can follow through the exhibits in the order you wish. Place posters and the pictures at eye level to attract people. If visitors will sit down to chat, consider what they can see from that level.

	Activity	Tasks	Dates
1	Focus Task Time	Get this clear. Clarify the task, create a working title, and agree the broad timeline. Read the assessment rules in the module handbook.	
2	Audience	Who – tutors, students, general public, employers, school students, and who else? … What is the audience background knowledge? Does the display need to be designed for one or many audiences?	
3	Facilities and space	Check table and poster board sizes. Is there a power point for computer or lights? Is there space for a game or quiz? Chairs encourage visitors to sit down and discuss issues.	
4	Draft Plan A Decide on the theme	Collect evidence, pictures, graphs … Be creative. Coordinate colours and fonts throughout. Create draft version of the display. Re-check assessment rules.	
5	Update Draft Plan Conversations	Critically review Plan A, collect further evidence, data, pictures to fill the gaps and revise the plan. Practise your presentation/conversations.	
6	Final production	Critically review the plan again, and the practice conversations. Finalise title, posters, stand materials, and interactive elements. Practise conversations again.	

Figure 23.4 Key steps to creating an exhibition stand

Where a stand is assessed, self-assess during the design stage using the assessment guidelines or Figure 23.5. Potential employers and external organisation representatives may be invited to the exhibition. Your motivation for producing a great display might include imagining you will talk to your first employer on the day. The following reflective quotes are from second-year undergraduate geographers after a two-hour exhibition for non-geographers.

The most important thing that the stand exercise helped me to learn is …
'Whilst talking to people new thoughts about skills and proof for them came

to you'. 'It is a skill to talk to people and to have to think on the spot, you have to adapt to different people and questions, you cannot just spurt out the same material'. 'I now know that I need to think through what I want to say in advance, so the message is really clear'. 'I learned a lot about interview techniques, so that when people fire difficult questions at me I can deflect them and try to answer them without looking visibly flustered – therefore very helpful'.

I most enjoyed … 'Creating the actual stand'. 'Getting lots of ideas from the other team members'. 'Seeing how good other people's presentations were'. 'Feeling a sense of achievement when someone enjoyed looking at the stand'.

I least enjoyed … 'Standing around waiting for people to talk to us, feeling like a spare part'. 'Being asked quite difficult questions because it was difficult to think on your feet and there was only limited time for research'. 'Some of the people were really difficult to talk to'. 'Talking to the first person, but it got better as time went on'.

Organising and running a stand is great fun and very exhausting. Work as a team and keep water and chocolate handy!

Stand Title ....
Group Names ....

Please grade on a 1–5 scale, where 1 is Useless, 3 is Average, 5 is Brilliant.	
1. How well did the group articulate the principal points and issues?	1 2 3 4 5
Comments	
2. How broad and effective was the evidence that justified the claims?	1 2 3 4 5
Comments	
3. How did individuals use examples and anecdotes to support their claims?	1 2 3 4 5
Comments	
4. How well did the stand materials support and enhance the points being made?	1 2 3 4 5
Comments	
5. How creative and impactful were the ideas?	1 2 3 4 5
Comments	
Additional feedback	

Figure 23.5 Assessment criteria for an exhibition stand

23.4 References and resources

BCUR 2014 Posters in Parliament, British Conference Undergraduate Research, http://www.bcur.org/posters-from-posters-in-parliament-2014/ Accessed 16 January 2019.

Hill J, West H and Kneale PE (2017a) Preparing for multidisciplinary undergraduate research conferences, *Journal of Geography in Higher Education*, 42, 1, 148–156.

Hill J, West H and Kneale PE (2017b) Making the most of multidisciplinary undergraduate research conferences, *Journal of Geography in Higher Education*, 42, 2, 311–316.

Macaulay Institute, 2010 Soils and Vegetation posters, www.macaulay.ac.uk/ soilposters/index.php Accessed 16 January 2019.

Purrington C 2017 Designing scientific posters, https://colinpurrington.com/tips/ poster-design Accessed 16 January 2019.

University of Liverpool, Making an impact with your poster, https://www. liverpool.ac.uk/media/livacuk/computingservices/printing/making-an-impact-with-your-poster.pdf Accessed 16 January 2019.

University of Texas, 2018 Poster Samples, Texas Undergraduate Studies, https:// ugs.utexas.edu/our/poster/samples Accessed 16 January 2019.

University of Toronto, Posters and Presentations, https://guides.library.utoronto. ca/posters_presentations Accessed 16 January 2019.

Relaxed maths 3

We do maths all the time; without using a calculator solve the following:

64 + 36 =	192 − 16 =	92 − 17 =
37 × 6 =	55 × 12 =	54 ÷ 6 =
817 − 77 =	665 − 33 =	162 × 4 =
448 ÷ 8 =	27 ÷ 3 =	76 + 153 =

Answers in Chapter 28.

Examinations and revision

Examinations should not be a journey into hitherto uncharted regions.

Exams are designed to test your understanding of interrelated material from coursework, personal research and reading. Your starting point is understanding the different types of examination paper, because your revision style will alter depending on the type of assessment. Keep in mind what examiners are seeking (Figure 24.1) so you can plan revision time effectively. Reflect on your revision style. A last-minute 'cramming' style is a high school concept. Taking the route:

Get the information → Learn the detail → Regurgitate in exam

leads to superficial, surface learning where information is forgotten relatively quickly. University modules build on the information from previous terms and years, with staff expecting you to put together ideas from these earlier modules, so revision needs to be wide-ranging.

Examiners look for:	Do	Don't
Understanding	Show that you understood the question. Keep answers focused. References please.	Don't ramble.
Relevance	Stick to the issues and the points that the question raises.	Don't get sidetracked. No irrelevant examples.
Analytical ability	Show that you understand the meaning of the question and can argue your way through points in a logical order evidenced with *references*.	Don't be disorganised.
Expression	Clear and concise writing. Make your ideas and arguments transparent to the marker. Maps and diagrams help enormously. A well-structured short answer will get better marks than a long irrelevant answer.	The length of an answer is no guide to its effectiveness or relevance, no rambling.

Figure 24.1 Helping examiners give you good marks

24.1 Examination essays

The minimalist advice here is to reread the advice on constructing academically evidenced arguments (Chapter 12), and essays (Chapter 18) and to write fast. Essays allow you to develop lines of argument, draw in diverse ideas and demonstrate your skills in argument, analysis, synthesis, evaluation and written communication. Remember to keep the academic content – geology, management, economics, petrology, biochemistry, hydrology, legal, geography, science, philosophy … – high, use evidence (many *references* and *case examples*) to support your arguments and cite the *references*. Good answers communicate enthusiasm for the material, avoiding boring repetition. Persuade the reader to agree with your ideas.

> **Essay plan lite**
> ✓ Introduction: background, context, framework
> ✓ Ideas, examples, references
> ✓ Ideas, examples, references
> ✓ Ideas, examples, references
> ✓ Interpretation and discussion
> ✓ Conclusions

⇨ TOP TIPS

→ If all questions look impossible, choose one where you have examples to quote, or the longest question. Long questions give more clues to structure an answer. 'Discuss the role of pressure, mineralogical crystal structure, temperature and other variables on radioactive decay in three contrasting rocks', is a question with loads of clues and parts to answer, whereas 'Discuss radioactive decay processes' is essentially the same question. It could be answered with the same information, but will be a minefield without a self-imposed structure.

→ Always plan your answer. Underline key words in the question, like <u>Discuss</u>, or <u>Compare and contrast</u>, and note the spatial and temporal scales to cover (Figure 24.2). Don't restrict examples to one place or country if the question covers global

It was two years before I realised you were expected to put references in exam answers, with dates. Humph.

or international issues. Do a quick list, mental map or spider diagram of the main points, and note ideas for the introduction and conclusion. Then rank the points to get a batting order for the sections.

→ On a three-question paper, plan Questions 1, 2 and 3 before writing the answer to Question 1. Your brain will run in background mode on ideas for Questions 2 and 3 as you write the first answer.

→ Watch the time. Leave a couple of minutes at the end to check each answer. Amend spelling, add extra points, references, and titles, scales and units to diagrams and maps.

→ Organise points, one per paragraph, in a logical order. Set the scene in your first paragraph and signpost the layout of the answer.

→ Be specific, especially about time and place: 'Early on …', is better expressed as 'From 1630 to 1720 …' or 'In the Jurassic era …'. The general, as in 'In UK towns we see …', is better expressed as 'In Winchester, Reading and Shrewsbury we see …'.

→ Put an interesting point in the final paragraph.

Typical assessors' comments in Figure 24.3 suggest that examiners only give marks when you explain in full. Clarify terms, answer all parts of the question, include many examples. Geographers need to include geographical cases and examples, and to relate geographical theory to the real world for high marks. Geologists often give very good technical answers, for example about chemical processes, but for high marks must explain where and why these chemical processes are relevant in geology.

Use technical language
- Avoid contractions: it's, let's, doesn't, wouldn't …
- Put in units, formulae, equations
- Add diagrams, graphs, data
- Avoid colloquialisms.

~~Cross things out~~, *very tidily.*

1:1	Hill slope	Household	Eon	mm	cm^2
1:50	Channel	Ward	Era	cm	m^2
1:1,250	River corridor	Town	Period	m	hectare
1:10,000	Flood plain	County	Epoch	km	km^2
1:50,000	Catchment	Region	Age		
1:1,000,000		Nation			
Truro	Parish Council	Second	By-Laws		
Cornwall	Local Authority	Minute	National Law		
South-west England	Regional Authority	Day	International Law		
United Kingdom	National Legislature	Year			
Europe	International Groups	Decade	Genetic diversity		
		Century	Species diversity		
		Millennium	Ecological diversity		

Figure 24.2 Spatial, temporal and other scales

Does not address the question	25%
Swarming with factual errors	35%
Fine answer to a question about coasts, pity the question was about karst	38%
Very weak effort– no attempt to explain examples or definitions	42%
Somewhat confused but some relevant points	42%
Good background but little analysis on the question set	44%
A scrappy answer with some good points	45%
Started fine then got repetitive	49%
Decent attempt, some examples but not focused on the question	50%
Not sufficient explanation/evaluation, you list some ideas at the end but these needed developing to raise the mark	52%
Reasonable answer but misses out the crucial element of… No equations, no science	54%
Has clearly done some reading but fails to write down the basics	54%
Reasonable effort as far as it goes, but does not define/explain technical terms – no examples, no references	55%
Accurate but generally descriptive, never really got to the 'evaluate' section	55%
Needs to learn about paragraphs, then fill them with organised content	55%
Reasonable effort – covers many relevant points, structure adequate but you need to organise points in shorter paragraphs	57%
Too much description – not enough argument	58%
Good – but needs to be more focused to key points	58%
Good introduction and well argued throughout, although with no evidence of reading. Good use of examples but unfortunately only from class material	64%
Has done some reading and thinking, lacks depth and references	65%
Quite good – a general answer, omits to define land-use but lots of non-lecture examples	64%
Excellent, comprehensive evaluation with new material	75%
Outstanding, original research as well as new reading and some original thinking	85%

Figure 24.3 Markers' comments

24.2 Short-answer questions

Short-answer questions search for evidence of understanding through factual, knowledge-based answers and the ability to reason and draw inferences. For short answers, a reasoned, paragraph answer is required. If the question starts 'Give four explanations for …', ensure you make four, fact-rich points. Take care to answer the question that is set! Add diagrams, graphs, equations and formulae and references as appropriate. See Chapter 24.9 for revision ideas.

> When did you last use a pen to write non-stop for two hours?
>
> Until all exams are typed online, **practicse writing with a pen**.

24.3 MCQs

Multiple choice questions (MCQs) test a wide range of topics in a short time. Modules with MCQs usually include practice sessions; be there. Make sure you understand the style of questions and the rules. MCQs may be used for revision, in a module test where the marks do not count, or as a part of module assessment where the marks matter. Lecturers design class tests very carefully to let you appreciate what you already understand, and where more research, thinking and revision will be helpful. In a final assessment, watch the rules.

Did you know all the MCQ questions and answers are scrambled in the test? We all did the same 50 questions, but they come up randomly. Trying to compare afterwards was hilarious. Matt wanted to know what was the answer to question 4, but we all had a different question 4.

Look carefully at the instructions which will remind you of the rules, such as:

- *There is/is not negative marking.* (With negative marking you lose marks for getting it wrong.) Or,
- *One or more answers may be correct, select all the correct answers.* (These can get you 100 marks on a paper with 60 questions.)

Questions come in different styles:

The 'Trivial Pursuit', factual style

These test recall of facts and understanding of theories, usually a small proportion of the questions:

1 CPRE stands for:
   a) Centre for the Promotion of Regional Excellence
   b) Capital Protected Rate of Exchange
   c) Campaign to Protect Rural England
   d) Center Parcs Rafting Exercise

2 Darcy's Law is usually expressed as:
   a) $Q = CIA$
   b) $Q = KIA$
   c) $Q = K(h/l)A$
   d) $Q = CID$

In this last example there are two correct answers, b) and c), and both should be indicated for full marks.

3 Which of the following was not a super continent?
   a) Pangaea
   b) Laurasia
   c) Davros Skaro
   d) Kenorland

Reasoning and application style

Reasoning from previous knowledge gives rise to questions similar to:

Which of the following sequences correctly rank air pollution emissions in the UK in 2010 (lowest emission first)?

a) Nitrogen oxides, Sulphur dioxide, Black smoke, Carbon monoxide
b) Black smoke, Nitrogen oxides, Carbon monoxide, Sulphur dioxide
c) Black smoke, Nitrogen oxides, Sulphur dioxide, Carbon monoxide
d) Black smoke, Sulphur dioxide, Carbon monoxide, Nitrogen oxides

Some questions give a paragraph of information and possible responses. You apply theories or knowledge to choose the right response or combination of responses. Some tests combine information from more than one course, as here, where a second-year soil analysis paper requires information from first- and second-year statistics modules:

A soil survey of 25 fields yielded measurements of NO_3, NH_4, P, K, particle size, OM, and bulk density, crop yield (tonnes per acre). The fields were planted with 4 crops (8 of barley, 8 of broccoli, 2 of sugar beet and 7 of broad beans).

1. Which statistical tests <u>might be used</u> to analyse the dataset comparing production of all 4 crops? Mark all the correct answers:
   a) Mann Whitney U, b) Student's t-test, c) Correlation, d) Multiple regression, e) One-way Analysis of Variance, f) Histograms, g) Chi-square, h) Nearest Neighbour analysis, i) Kruksal-Wallis test, j) Two-way Analysis of Variance.
2. Which statistical tests <u>should not</u> be used to compare broccoli and sugar beet yields? Mark all those <u>not</u> correct:
   a) Mann Whitney U, b) Correlation, c) One-way Analysis of Variance, d) Histograms, e) Chi-square, f) Nearest Neighbour analysis, g) Kruksal-Wallis test, h) Two-way Analysis of Variance, i) Student's t-test, j) Multiple regression.

Data response style

A combination of maps, diagrams, data matrices and written material are presented, and a series of questions explore possible geographical, environmental and geological interpretations.

Appropriate revision techniques for MCQ papers include creating quiz questions, which forces you to concentrate on details, and is great fun. Use **Try This 24.3** and **Try This 24.4** for practice.

24.4 Laboratory examinations

In addition to coursework, some physical laboratory and computer courses have examinations to test that you understand the applications of the experiments and practical work. Revise with past papers, or devise new questions. Try variations on:

- *If soils experiment … was rerun with soil samples from Exmoor/Tunisia/Spitsbergen how would you expect the results to change?* Essentially a 'what will happen if the samples come from different environments' question. They are designed to test understanding of controlling and varying factors.
- *The flume tests were rerun with flows of … m.s^{-1}. How would you anticipate the sediment erosion and deposition patterns would alter from those observed in the initial experiments?* This is another 'what if we change the parameters' question that asks for logical speculation on outcomes.
- *After Experiment 12 was completed the … meter was found to have an error of 15%. Explain how this information affects the interpretation and how it can be accounted for in the calculations.* A 'how can you cope with instrument and operator error' question.
- *Explain where errors can arise in … (water sampling, pollen counting …) and the impact they have on data analysis and interpretation.* A general question covering data and analysis errors.
- *Describe the safety implications involved in …, and explain how they can be minimised.*
- *Explain how and why instruments to measure … should be calibrated before field use.*
- *Outline the field sampling and laboratory tests you would use to explore … How accurate would you expect your results to be?* Explores the approaches and tests you would use, and their relative accuracy.

24.5 Oral examinations – vivas

Oral examinations, sometimes called vivas, are increasingly rare but in some degree schemes, in some universities they are part of a module. This is an exciting opportunity to discuss a topic or project. Think of it as a presentation (Chapter 22) without the presentation: you move straight to the Q&A session. This helpful skill should be more widely practised because you are much more likely to 'explain' an idea or project than write about it at work.

DON'T PANIC! RE-read your project report and course materials, get a good night's sleep, avoid unusual stimulants (always sound advice) and wear something tidy. Examiners are interested in your brain not your wardrobe. Celebrate later. (See Chapter 25 for the dissertation viva.)

24.6 Take-away or take-home examinations

Papers written outside an examination room need the same revision and preparation. Know exactly when the title/paper is being released, and be ready. Make sure your diary is clear, food is organised, and you have a space to concentrate and to write without interruptions.

You may have between one and three days to answer the question. This allows you to demonstrate your research skills, ability to write a coherent argument and present a professional document. It's an authentic assessment because it's a typical workplace activity.

Get the notes together in advance, ensure access to library resources and get your brain in gear by thinking of possible essay topics. Do not leave it all to the last minute.

This type of assessment demonstrates your ability to research and write in a short timeframe. Mention it in job applications.

24.7 Revision

Finding ways to accumulate and consolidate the information will be much more helpful than cramming at the last minute. In life (most jobs) and for effective study you accumulate knowledge and apply it continuously. The most effective revision results from reading and reviewing material as an ongoing process, built into your weekly timetable, because it improves the amount of detail you remember. Normal learning activities reap rewards at exam time because they are also 'revision' activities that reinforce your learning all the time. The following will help:

✓ Get your brain in gear before a lecture by reading last week's notes.
✓ Read and reread all discussion notes. Read new things. Talk!
✓ When tired, use your computer's Read Aloud (Texthelp 2019) speech function to hear your notes, also useful when cooking, running and in the bath.
✓ Use SQ3R in reading and making connections to other modules (**Try This 10.2**).
✓ Think actively around issues (Chapter 8), ask questions as you research.
✓ Check that you have good arguments (Chapter 12), spot gaps in the evidence and use that information to determine what you read next.
✓ Keep fit, take breaks, eat well.
✓ Timetable 'revision slots' to continue these learning processes.

Our best idea was the study buddy group. Explaining something to the others made me understand the points and remember the details.

24.8 Revision – getting organised!

I finish five modules this week, have three term papers to hand in on Friday, and the first exam is on Monday. So when do I revise?

Tight deadlines are part of university life. Tutors will be emphatically unsympathetic about complaints that five essays are due on the same day with examinations two days later. Sorting out your schedule is your problem; exercise all your time-management skills (Chapter 3). Lecturers know that when they arrange essay deadlines earlier, to reduce the end of term deadline crunch, they get even more complaints because '*the deadline's too early*'. (It's a no win situation for staff.)

Try to assign 30 minutes every week to review and reflect, to think around issues for each module, and decide what to read next – this is **revising**. It is an ongoing process, and you will be well ahead of the majority! As exams get closer, sort out a work and learning timetable, say for the five weeks before exams. Put every essay, report and presentation deadline on it, add the exam times, lectures, tutorials and all your other commitments, and have a little panic. Then decide that panicking wastes time and sort out **the plan.** You can do six essays and revise 12 topics and be social and sporty to relax. Block the time in two-hour slots, with breaks to cook, shower and eat (maybe not in that order – see Chapter 3). Use the time on your tablet/watch or computer. See how that works – for some people 90-minute slots are better. Then organise group revision, discussion and sharing opportunities with friends, GET STARTED.

> *I have a wall chart with the weeks and exam dates. Each module has different coloured Post-its. Things to do go on post-its. I move them around, it's easy to see where I spend time on each module.*

➪ TOP TIPS

→ Start earlier than you think you need to!
→ Put revision time into your weekly plan and use it to think.
→ Keep a revision record; attempt to allocate equal time to each paper.
→ Speaking an idea aloud or writing it down, lodges information in the brain more securely than reading.
→ Think actively around issues (Chapter 8), ask questions as you research.
→ Keep fit, take breaks, eat regularly and well (remember fruit and veg).

Revising is not something you must do alone. Think with study buddies. Group revision will:

☺ generate comments on ideas and add other people's perceptions to your brain bank;

☺ let two or three brains disentangle theories, explain decentralisation, drift or drumlins;

☺ be more fun (less depressing?);

☺ make you feel less anxious, more self-confident;

☺ show where you need extra study and where you are already confidently fluent with the material.

> *Always take three pens into the exam. You never know when one will get fed up and stop writing.*

Don't be put off by someone claiming to know it all because they have read something you have not (no one reads everything); if it's so good, ask for an explanation.

How do you revise?

Revision is where you put the puzzle together.

What were your thoughts as you left your most recent exams? Ignoring the obvious 'Where's the bar?' jot down a few thoughts and then look at **Try This 24.1**. This is feeding back to yourself so you can plan effectively.

TRY THIS 24.1 – Post-examination reflections

Jot down thoughts about your last examinations, then look at these post-examination thoughts of rather jaundiced first-year students. Tick the points you empathise with; what else concerns you? Make a plan for next time and act on it.

☹ *No practice at timed essays since 2017; I'd forgotten how after eighteen months.*

☹ *I ran out of time to write.*

☹ *A lot of the lecture detail seemed unnecessary for the questions.*

☹ *There were too many specific places and dates to remember.*

☹ *Lack of revision.*

☹ *I didn't know what was relevant.*

☹ *Most of the time I was thinking about going to the bar after.*

☹ *I needed more direction towards the questions.*

☹ *You couldn't learn the whole course; I picked the bits that were missed.*

☹ *There were loads of facts in the lectures, but nothing to apply it to.*

☹ *I mostly relied on the stuff we did at school and made some things up.*

☹ *I had all this stuff about Calcutta and he asked about Brazil.*

☹ *I think I messed up because the detail in the lectures was confusing and the big handout added more detail that wasn't in the lecture. I didn't really get started on it because there was too much to tackle.*

24.9 Revision action

Revise actively

Sitting in a big armchair in a warm room reading old notes or a book will send you to sleep! TRY some of these ideas:

☺ The rubrics for each exam paper will be somewhere (module handbook, VLE, library). Knowing the style, number to answer, and the marks available for each question will help you focus your revision.

☺ Look at the feedback you have had from coursework, tutorials, the VLE …

> *Making notes from notes is not fun. Recording notes as a podcast and hearing it later was more fun.*

☺ Make summary notes, ideas maps, lists of main points and summaries of cases or examples.

☺ Draw and redraw sections, maps, diagrams and graphs.

☺ Sort out the general principles and learn them.

☺ Look for links, ask questions like:
- Where does this fit into this course/essay/other modules?
- Is it a critical idea/detailed example/extra example to support the case?
- Is this the main idea/an irrelevance?

☹ Write outline answers, then apply the criteria of Understanding, Relevance, Analysis and Expression. Ask yourself 'Does this answer work?' 'Where can it be improved?'

☹ Practise writing an answer in the right time. Sunday morning is a good time! Put all the books away, set the alarm clock and do a past paper in the right time. DON'T PANIC – the next time will be better. Examination writing skills get rusty, they need oiling before the first examination, or your first answers will suffer.

☹ Remember you need a break of five minutes every hour or so. Set the timer function on your phone/pad/computer/cooker.

> When did you last use a pen to write non-stop for two hours?
>
> Until all exams are typed online, **practise writing with a pen**.

☹ Plan to swim, walk or visit the gym for a short period each day; oxygen revitalises the brain cells.

☹ Use the Active Reading SQ3R technique (**Try This 10.2**, p 124) to condense notes to important points.

☹ Aim to review all your course notes a week before the exams! You will panic less, have time to be ill, and feel more relaxed. (OK, OK, I said AIM!)

☹ Staying up all night to study is not one of your best ideas in life.

Reviewing the first four weeks of a module should, ideally (ho-hum), be completed by weeks six to eight, so that the ideas accumulate in your mind. The rest of the module will progress better because you understand the background and there is time to give attention to topics in the last sessions of the module. Can you be 'ideal' for one module, maybe the really difficult module?

Outline answers

Outline answers are an efficient revision alternative to writing a full essay. Write the introduction and conclusion paragraphs in full, NO CHEATING. For the central section, draft one-sentence summaries for each paragraph, with *references*. Add diagrams, maps and equations in full, practice makes them memorable. Then look carefully at the structure and balance of the essay. Do you need more examples? Is the argument in the most logical order?

Key words

Examiners tend to use a number of key words to start or finish questions. Words like Discuss, Evaluate, Assess, Compare and Illustrate. They all require slightly different approaches in the answer. Use the **Try This 24.2** game to replace key words and revise essay answers.

⚙ TRY THIS 24.2 – Replacing key words in examination questions

Take a question, any question from a past examination paper. Do an outline answer structure. Then replace the key words with one or more of the words below. Each word changes the emphasis of the answer, and the style and presentation of evidence. 'Describe the impact of …' and 'Evaluate the impact of …' need different styles of answer. (Key words and meanings adapted from Rowntree 1988.)

Analyse	Describe, examine and criticise all aspects of the question, in detail.
Argue	Make the case using evidence for and against a point of view.
Assess	Weigh up and make a judgement about the extent to which the conditions in the statement are fulfilled.
Comment	Express an opinion on, not necessarily a long one, BUT often used by examiners when they mean Describe, Analyse or Assess (cover your back).

Compare	Examine the similarities and differences between two or more ideas, theories, objects, processes etc.
Contrast	Point out the differences between … Could add some similarities.
Criticise	Discuss the supporting and opposing arguments for, make judgements about, show where errors arise in … Use examples.
Define	Give the precise meaning of … Or, Show clearly the outlines of …
Describe	Give a detailed account of …
Discuss	Argue the case for and against the proposition, a detailed answer. Try to develop a definite conclusion or point of view. See Comment!
Evaluate	Appraise, again with supporting and opposing arguments to give a balanced view of … Look to find the value of …
Explain	Give a clear, intelligible explanation of … Needs a detailed but precise answer.
Identify	Pick out the important features of … and explain why you made this selection.
Illustrate	Make your points with examples or expand on an idea with examples. Generally one detailed example and a number of briefly described, relevant supporting examples make the better argument.
Interpret	Using personal experience, explain what is meant by …
Justify	Give reasons why … Show why this is the case … Argue the case.
List	Make a list of (usually means a short/brief response), notes or bullets may be OK.
Outline	Give a general summary or description showing how elements interrelate.
Prove	Present the evidence that clearly makes an unarguable case.
Relate	Describe or tell a story; see Explain, Compare and Contrast.
Show	Reveal in a logical sequence why … See Explain.
State	Explain in plain language and in detail the main points.
Summarise	Make a brief statement of the main points, ignore excess detail.
Trace	Explain stage by stage … A logical sequence answer.
Verify	Show the statement to be true; the expectation is that you will provide the justification to confirm the statement.

The quiz approach

Devising revision quizzes can be effective. Use **Try This 24.3.** You do not need to dream up multiple answers, use the question style to focus your draft answers.

> **TRY THIS 24.3 – Designing quiz questions**
>
> The first examples need factual answers good for MCQs and short answers. The second set need the same factual knowledge but a more reasoned and extended response. They are useful for revising short-answer questions and essay paragraphs.
>
> **Factual:** for MCQs, short answers, and checking you have facts ready. Precise WHAT WHICH WHERE questions; for example:
>   Which nutrients are the main causes of eutrophication?
>   Where do the nutrients that cause eutrophication principally originate from?
>   What were the impacts of eutrophication in Lake Victoria?
>   What is the average water supply and sanitation coverage in mega-cities across sub-Saharan Africa?
>   What do … (authors) conclude is the most significant cause of eutrophication?
> **Exploratory:** for essays, WHY and more general questions that let you explore evidence and examples:
>   Why do wetlands act as buffers to eutrophication?
>   What does eutrophication mean in practice?
>   What are the implications of not treating wastewater and sewage from urban areas in sub-Saharan Africa?
>   What are the long-term effects of wastewater disposal to ground water?
>   When and why do some fish species increase their numbers and others become extinct following eutrophication?

To get into the swing, use **Try This 24.4** and **Try This 24.5.** The first asks for quiz questions based on your notes from lectures and additional reading, and the second uses the same technique with a journal article.

> **TRY THIS 24.4 – Quiz questions from notes**
>
> Pick a set of notes and create a short quiz, ten questions, in each of the two styles. First, short factual questions, then more extended questions. Write the questions on one side of the page and answers on the reverse, so you can test yourself (without cheating too much).

> **TRY THIS 24.5 – Quiz questions from a journal**
>
> Pick an article from any reading list. Devise a set of five short and five extended questions that explore the topic. Write the questions on one side of the page and answers on the reverse, test yourself next week.

Revision is an opportunity to put ideas together, looking for explanations and insights. It's a creative process with ideas gelling and developing as you reread and reconsider. Look at new material, make sure you understand definitions, consider timescales and spatial scales. Note-making is an integral part of revision. Some general questions running in your head will encourage a questioning, active approach to revision, see **Try This 24.6.**

TRY THIS 24.6 – Good revision questions

Make a list of the starting phrases that could be useful when compiling revision questions and quizzes. For example:

1. What is the purpose of …? (cryosphere, trilobite, public–private partnerships)

2. Why is … an inadequate explanation of …?

There are more suggestions at the end of the chapter, but have a go before looking.

Creating beautiful revision timetables is prevarication at best. One timetable is enough.
Just start revising.

24.10 Exam day

Be practical: Relax, you did all that reflection and reviewing; the examinations will be most agreeable. Check and double-check the examination timetable and room locations – they can change. Plan to arrive 20 minutes early to find your seat number, visit the loo and relax. The student handbook explains what to do if you are delayed. Make sure you have your ID, pen, spare pens, calculator …, whatever is required. For online exams you need your ID and passwords. Read the information on the exam paper (rubric), and listen to the invigilator. If there is a problem with the paper or computer tell an invigilator at once.

Use the time in proportion. For example, on a two-hour paper that has six short questions and one essay there will probably be 50 per cent for the essay and 50 per cent split equally between the six short questions. One hour for the essay and nine minutes for each short question, leaving six minutes to read the paper and plan the answers. Answer the number of questions required, no more and preferably no less. Leave time to do justice to each question. Don't leave your potentially best answer until the end. Equally, don't spend so much time on it that you skimp on the rest. Make sure there are no questions on other pages.

Should you be seized with anxiety and your brain freezes over, use the 'free-association', brainstorming approach. Write out the question and then look at each word in turn, scribbling down the first words that occur to you, authors' names, examples, related words … This should generate calm and facts. Plan from the spider diagram you have created.

After every exam review your answers as soon as you can. What did you learn from the exam process (feedback to yourself)? University exams are a little different to school ones; reflect on (think around) your revision and exam technique, and what you might do next time. Questions that might occur to you include:

? Had I done enough revision? (The answer is usually no.) Where could I have squeezed in a little more revision?

? Was there enough detail and evidence in each answer to make an excellent case?

? What do I do next time? What revision activities would be useful?

24.11 Exam anxiety

Being anxious beforehand and on the day is normal. Most people are feeling nervous, sick, perspiring, dashing to the bathroom and worrying. Revision should reduce your anxiety because you've taken the time to become very familiar with the relevant information, and practised writing draft answers. If you start panicking, stop, breathe slowly and relax.

Having a balance of activities while revising is important (Chapters 3 and 5). Make a revision plan that includes time for sleeping, eating and socialising as well. Getting some exercise will help you sleep. Sleep is really important, so your brain has time to absorb and organise all the information you've captured during the day. A regular and prompt bed-time is good. Keep artificial stimulants for later.

It's easy to assume everyone else is okay, whereas in fact everybody is emotional and that changes their behaviour. Keep calm, no one's perfect, and remember that millions of students pass their university exams every year.

Your university probably provides additional support to help students prepare for exams and manage the associated stress. Look for resources in your student learning centre, counselling or library.

24.12 Exam feedback

There are many ways students get feedback on their exam performance. Find out what your School does and get involved. There may be:

➤ A day when all examiners are available with all the papers. You go through what you did well and where you can improve.

➤ A tutor who asks you to come for a chat.

➤ Good answers posted on the VLE with comments on why this was an effective answer.

➤ A podcast of the marker's comments for individuals or for the whole class.

➤ Comments sent to your email account.

The trick is to use the feedback effectively. You cannot change the mark, the module is over, but turning up and talking with a tutor will give you insights into your revision process and examination technique, which will help you next time. Useful feedback improves your skills.

24.13 References and resources

Hay I 2002 Coping with exams: dealing with the cruel and unusual, in Rogers A and Viles H (Eds.) *The Student's Companion to Geography*, (2nd Edn.), Blackwell, Oxford, 190–196.

Knight PG and Parsons T 2003 *How to do Your Essays, Exams and Coursework in Geography and Related Disciplines*, (2nd Edn.), Routledge, London.

Learnhigher 2012a Academic writing for exams podcast, http://www.learnhigher. ac.uk/writing-for-university/academic-writing/academic-writing-for-exams-podcast/ Accessed 16 January 2019.

Learnhigher 2012b Memory and exams, http://www.learnhigher.ac.uk/learning-at-university/assessment/memory-and-exams/ Accessed 16 January 2019.

Learnhigher 2012c Revising for maths assessments, http://www.learnhigher.ac.uk/learning-at-university/assessment/revising-for-maths-assessments/ Accessed 16 January 2019.

Learnhigher 2012d Revision techniques podcast, http://www.learnhigher.ac.uk/writing-for-university/academic-writing/revision-techniques-podcast/ Accessed 16 January 2019.

Rowntree D 1988 *Learn How to Study: a guide for students of all ages*, (3rd Edn.), Warner Books, London.

Texthelp 2019 Boost Achievement, https://training.texthelp.com Accessed 16 January 2019.

University of Leicester 2018 Revision and exam skills, Student Learning Development, https://www2.le.ac.uk/offices/ld/resources/study/revision-exam Accessed 16 January 2019.

University of Melbourne 2018 Revision and exam preparation, https://students.unimelb.edu.au/skills/revision Accessed 16 January 2019.

Wordsearch 4

Can you find the words and phrases in the grid? Words are written forwards, backwards and diagonally. Solution in Chapter 28.

Acids, areator, alluvium, basalt, carat, carbon, carbonate, cloud, endangered, species, energy, fjord, fumerole, gabbro, hematite, hydrocarbon, karst, lithosphere, metamorphism, marble, mercury, ocean, oxbow, oxide, parasite, pesticide, radiation, recycled, salinity, stack, swarm, wildlife.

```
E E H T S R A K O B W D N N C
N B M E A I N C A X U O O A A
E D S S M R E S D O I B B O R
T E I A E A A B L T R D C X B
I L H L N L T C A A S G E S O
S C P I T E D I C I T S E P N
A Y R N L L D O T A A M W E A
R C O I M A R B L E C U I C T
A E M T R D T E D E K I L I E
P R A Y Y G A B B R O V D E N
L I T H O S P H E R E U L S O
N D E R E G N A D N E L I W B
E O M E L O R E M U F L F A R
D R O J F R E A R O T A E R A
A L E N E R G Y R U C R E M C
```

Answers: Try This 24.6 – Good revision questions

1. Who are the three main authors to quote for this topic?
2. Name two examples not in the course text or lectures for …
3. What are the main characteristics of …?
4. Outline the relationship between … and …
5. What are the limitations of the … approach? (core-periphery perspective, nebular hypothesis, ice-cores as climate records)
6. What methodology is employed to …?
7. At what point does this process become a hazard?/of concern? (algal blooms, monopolies)
8. How has human impact affected …?
9. What was the main aim of …? (open-ended questions, reconstructing past climates)
10. What is the spatial scale involved in …?
11. How is … calculated? (isotope decay, Reynolds number, market potential)
12. Outline the sequence of events involved in …

13. What are the implications of … not occurring? (fire in ecosystems, inward investment)
14. How has … adapted to …? (vegetation – arid environments; housing tenure – post-war Britain)
15. Outline the different approaches that can be taken to the study of …
16. Will this work in the same way at a larger/smaller scale?
17. How influential has … been?
18. What are the … (regional, national, cultural) … implications of …?
19. What is meant by …? (commensalism, regional metamorphism, logical positivism, crude death rate, safety factor)
20. Why is … important for …? (copper – plant growth; secondary census data – population studies)

25

Introducing dissertations and extended projects

The First Law of Dissertations: Anifink tht kin goo rongg, wyll.

Most UK degree courses have a final-year module that involves personal research and an extended piece of writing. Whether it is called a dissertation, project, long project or …, it will count for 20–60 credits and therefore can have a considerable impact on your final degree result. This chapter does not in any way replace or pre-empt your department's guidelines and advice. It just aims to answer questions typically asked by students in the first two years of a degree, when a dissertation can be viewed as an academic Everest to be assaulted without aid of crampons or oxygen, and it mentions timescales for effective planning. Each department has its own timing, style, preparation, expectations of length and monitoring procedures. Most departments produce detailed guidelines and run briefing sessions so everyone is aware of the rules. Missing these briefings is a BAD idea; one of your better ideas will be to reread the module handbook/web page every six weeks, to remind yourself of expectations, milestones and guidelines. More detailed information will be found in Parsons (2015) and Peters (2017). Considering going abroad? Start with Nash (2000a, b) and Scheyvens (2014).

Skills addressed during dissertation research include autonomous working, setting and meeting personal targets in research, professional report production, problem solving, and many more. Compile a 'skills used' list while researching and cite these in job applications and interviews.

25.1 A dissertation is not an essay, what is the difference?

Essentially, a dissertation is an opportunity to enquire systematically into a topic or problem that interests you, and to report the findings for the benefit of the next person to explore that material. Some dissertations

> *It was amazing, if v scary to do just what I wanted, and to go to Namibia too.*

are published, see Geoverse (2018), Reinvention (2018) and Plymouth Student Scientist (2018). Aiming to produce a product that is worthy of publication is appropriate. It is an opportunity to explore material, develop an idea, conduct experiments, analyse information and draw mature conclusions from the results. The results need not be mind-blowing – discovering a new continent or geological era is rare, and the world will not end if your research has a completely expected or completely unexpected answer. Report what you find.

Giving yourself time is vital. Most dissertations start in second year, providing ample thinking time and opportunities for pilot work (Figure 25.1).

On some programmes students either research an academic topic presented as a traditional research dissertation, or undertake a project working for an external organisation. This type of workplace or work-based project is likely to involve producing a report for the organisation or employer, and a reflection on your process of working to complete the research. The style and content for these reports will vary considerably depending on the nature of the project and the negotiated requirements of the hosting organisation. This type of workplace project provides an excellent way of gaining employment experience, building content for your CV and evidence for interviews. It is particularly important to check your university module submission requirements so you meet the requirements of your degree programme as well as providing an excellent report for the employer.

Geologists and earth scientists: the independent mapping project may run alongside a dissertation; see the Geological Society (2017) web pages for useful introductory information.

Consulting your university supervisor throughout your placement project work is vital for making sure you focus on and include essential elements. Most of the points below matter for all styles of project.

25.2 When to start?

Typically, you will explore a topic by yourself, deciding when, where, how and in what detail to work. Supervisors will advise, provide clues as to what is expected, but basically it's up to you to plan and organise. Look at the handing-in date, say 1 May; now work backwards.

1. Allow three weeks for slippage, flu, visitors, Easter, career interviews, despondency (April).
2. Writing-up time: allow 3–4 weeks – OK, it should take 6–7 days, but you have other things to do in March (March).
3. Analysing the data, fighting computer systems, getting print-outs, creating graphs … (February).
4. Recovering from New Year, start-of-year exams, career interviews (January).
5. End-of-term balls, parties, preparing for Christmas, end-of-term exhaustion, Christmas, family (December).

6. Get data from the local authority, environment agency, census, field collection, laboratory analysis … allow 6–8 weeks (October and November).

7. Have an idea, check out the library, think, talk to supervisor, have two more ideas … settle on a topic, field visit, pilot survey, allow 6–18 weeks (start up to a year earlier).

A final-year project or dissertation, designed to be done between October and May, will consume chunks of time through the period. With an early October start date (meaning

> I got to Tunisia, then realised I had no clinometers.

late October, because weeks 1 and 2 are needed for recovering from the vacation, starting new modules, catching up on friends and partying) a 1 May finish is very close. Start planning in Year 2. Can fieldwork or data retrieval be completed in advance? Can you check out potential field sites in Easter Vac of Year 2? Summer vacation fieldwork, and especially fieldwork abroad, needs advanced planning, the right equipment, clear formulation of hypotheses, and a workable, planned timetable.

> I had this great idea, got to the field site and discovered they had built a housing estate all over my river. Right Muppet moment.

A dissertation should not be rushed if it is to get a high mark (Figure 25.1). Most low marks for dissertations are won by those who start very late, and those who leave the crucial thinking elements to the last week. You may have noticed useful thoughts occurring a couple of days after a discussion or meeting. You wake up thinking,

'*Why didn't I say …*' High marks become attached to third- rather than first-draft dissertations. This means being organised so that there is thinking time at the end.

Figure 25.1 The dissertation process

25.3 Choosing your topic

Aim to explore an idea or hypothesis that fits your subject interests *and* captures your imagination. If you are not interested, you are unlikely to be motivated to give enough time and thought. You may want to complete a dissertation that is relevant to your chosen area of employment. A dissertation may present an opportunity to indulge a hobby (mountain biking …), or to visit someone, BUT must address a good question, as in 'the impact of mountain biking on soil erosion', or 'immigrant labour in Toronto suburbs (where my aunt lives)'.

Questions that start: How …? To what extent …? Which factors cause …? help to focus thinking. Coming to judgements can be more difficult, as in 'Are Keynesian economic values more useful than …?' Avoid topics that are very big and largely unanswerable: 'Will geography have relevance in economic planning in the second half of the twenty-first century?' Steer clear of topics where data are impossible to obtain, for example 'A discussion of the geology and landscapes of Pangea at latitudes 25–45°N', or, 'An evaluation of the geomorphological history of the sixteenth moon of Saturn through analysis of the geochemistry of surface samples'.

 TOP TIPS

➜ Think small at the start.

Spotting gaps

Throughout the degree course, make notes of thoughts like: 'Why is it like that?' 'Is it really like that in the supermarket, town, glacier, desert, landslide, quarry that I know?' 'Is that really right?' 'But I thought …' Also, note when lecturers say, 'but this area is not researched', or 'this was investigated by x in 1936, no developments since'. Another entry point is when an argument seems to have gone from alpha to gamma without benefit of beta. Now, it may be that the lecturer does not have time to explore beta on the way, and there may be a great deal known. Alternatively, beta may be unexplored, a little black hole in need of your torch.

Having spotted an apparent 'hole', take a couple of hours (online and library) to research what is known. Many a tutor has sent a student to research a possible topic, to be greeted by the response that '*the topic is no good because there are no references available*'. SUCCESS IS FINDING LITTLE OR NOTHING. This is what you want: a topic that is relatively unexplored so that you can say something useful. Researching for an essay you want lots of definitive documentary evidence to support arguments. For a dissertation you want to find little, or contradictory material, to support the contention that this is a topic worth exploring. But if there is lots of literature, that's OK, use it as a framework, a point to leap off from, to explore and extend. Watch too for areas of the subject where the published literature goes out of date very quickly. Anything to do with government policy, such as housing, changes with a new party in power, and can change rapidly within the lifetime of an administration as policy evolves. You can investigate the current position as compared with last … (month, year, decade).

Browsing is a primary dissertation research technique. Immerse yourself in material that is both directly related and tangential to the topic. Wider reading adds to your perspective and provides insights into alternative approaches and techniques. You cannot use them all, but you can make the examiner aware that you know they exist.

Topics

There will always be many questions to be answered. One or two topics are done to death. There are fads that relate to whatever is taught in the two weeks before dissertation briefings, and standard chestnuts often expressed as, 'I want to do something on air quality; faulting; global warming; water chemistry changing downstream; dolphins; crime in the city; social housing'. These are quite positive statements compared to the student who wants to do 'something historical', 'something to do with weather' or 'something while canoeing in Scotland'. The choice of dissertation is *your* responsibility. There may be a departmental list of suggested topics, but thinking to decide on a valid research hypothesis is your business.

Following a personal interest might lead to an extremely good dissertation on 'badger habitats', 'the environmental impact of the Felphersham bypass' or 'the impacts of legislation on the fishing industry in the 2020s'. You will get lower marks if you bias your answer with anti-badger-baiting literature only, your diary from 'my summer as an eco-activist', or present an entirely pro-Greenpeace fishing story. Writing any of these stories would be fine in a newspaper article that seeks to put across a single viewpoint. A dissertation is an objective academic exercise and so requires objective reporting. Having extremely strong feelings might lead to an excellent or a direly unbalanced report.

Survey-based studies

Most GEES projects involve collecting and evaluating data. Hence most degree courses include modules on document analysis and survey techniques, including quantitative analysis, questionnaire design, observation and interview techniques. Results are reported through quantitative and qualitative summary of the responses, statistical analysis and modelling, as appropriate. Where data are required on a longer historical scale, secondary data, such as the census, national or government data, national and international statistics and library archives, are vital sources.

Collecting data has the advantage that you know where and when it was compiled, you have a feel for its accuracy and know exactly why you asked particular questions. The disadvantages include the time-consuming nature of repetitive sampling, and intelligent consideration of sample representativeness is required. The time available for data collection is short; data are often representative of one day, week or month in one season at a limited number of sites. This is better than no data and a perfectly good way to proceed, but be wary of over-generalising from the individual to the global case. Avoid inferences like 'The 30 grains of sand monitored in the first order tributary of the River Swale moved 6mm in 12 weeks this last summer, therefore we can conclude that River Nile sediments have moved an average 16 miles since the birth of Christ'.

The critical issue with secondary data is access. Online is great but if it is coming from an organisation then it will probably take longer to arrive than you hope. Can you discover why it was collected originally? The purpose of the original collector, and your purpose, are almost certainly different. This is not a problem, but you need a paragraph somewhere to tell the reader you understand how the original data collection, curation and collation processes impact on your results. Consider data decay issues: does your data characterise the period you are interested in, or is it X years old? It is tempting to match the best secondary data available with current observations and not to address the data age gap. It may not matter or it might be really important, depending on your topic. Ask whether the arbitrary nature of sampling influences results. For practical reasons, sampling is undertaken on an hourly, weekly, bi-monthly … timescale; does this introduce bias in the results?

Case studies

Exploring 'Social diversity in two villages in Westeros' or 'Managing flooding in Ankh-Morpork' or the 'Fault patterns on Gallifrey' can be great. Case studies allow detailed examination of a specific area, time period, incident, event, cross-section or computer simulation. The aim is to understand how the world operates by examining a typical example. Alternatively, you might want to explore the exceptional case to see how the general model responds in extreme circumstances. Watch out for conditions where your 'typical study' deviates from the average.

Time series or cross-sectional studies

Time series studies let you examine the way different groups respond to the same phenomena or to a changing situation. Studies of downstream, down slope, down fault … sediments, water quality, soils, and changing weather patterns over time are regular topics. Ensure the background control conditions are kept as constant as possible. It should be no surprise to discover bedload chemistry changes downstream if the river passes a number of geochemically distinct bedrocks, two mines and three sewage treatment works on the way.

You explore samples from time slices, for example the shopping patterns of people in each 10-year age cohort. By comparing behaviour and attitudes you can differentiate patterns of shopping in the under-10s, teenagers and older cohorts. Remember the influence on different cohorts of cultural and social factors.

Theory-based topics

An examination of the geographical, environmental or geology theory related to a part of the subject is perfectly acceptable and potentially very rewarding, though it may seem a little risky at first glance. What will not get high marks is a voyage into your personal opinions on 'the Weberian approach to social class', or a personal diatribe on the ecological impacts of mine waste tailings. Tackling a theory-based dissertation allows you to think through a series of related issues, but there must be evidence to support the argument. This requires careful inspection of information. Be especially careful to locate the supporting and detracting arguments and come to a balanced judgement considering the range of material. Set the theory in its discipline context; this requires wide reading around alternative positions.

Before embarking on a theoretical study, talk the ideas through with a tutor. You need enthusiastic advice and encouragement to develop your thinking. Having read and thought further, go back and discuss the new ideas with housemates, tutor group, staff and postgraduates. Share your ideas and become familiar with them. Browsing and discussion are essential elements of the theoretical geographer's research methodology.

Developing or evaluating a technique

This makes a good dissertation if careful attention is paid to sample size and getting the statistics right. You might look at the accuracy of an instrument in different environments; at variations on mathematical equations; at a sensitivity analysis of part of a simulation model; or at the comparative accuracy of measuring something with three or more techniques. The advantage is that the problem is defined fairly tightly, and provided enough measurements are made the precision of the results can be evaluated.

What if… project

This is a speculative dissertation, and requires sensible thought and valid forecasting frameworks. It must be a topic where you can gain some information, from data or historical expertise, to build the argument, and the counter-argument. 'An evaluation of the geographical consequences of re-nationalising the UK railway system in 2025' would be a speculative piece of research embedded in an examination of the historical consequences of the last nationalisation of the railways and commenting on issues surrounding privatisation. Reporting a number of 'what if' scenario forecasts using transport and economic models will ground the speculations and arguments. This type of dissertation demonstrates synthesis skills.

Beware

There is a temptation to feel you should show one thing or another, that the issues are black and white. Most environmental matters are complex, with subjective and objective disagreements. Much of what we know is an approximation; researchers hope to make the best inferences, given the information-gathering procedure available at the time. Interview-based information is subjective, relying on people telling the truth and there being time to collect enough material. Even objective data has its inaccuracies. Population geographers can supply you with data on a country's population, but they will also tell you that even where the census is well regulated, individuals avoid the count for personal reasons, and errors occur in collection and calculation procedures. The estimated historical population of England in 1777 is just that, for 977 and 477 'population facts' are debatable estimates. When exactly did the Holocene start, never mind the Cretaceous? Knowledge is partial; go for your best interpretation on what you know now, but keep looking for and recognise areas of doubt in your arguments.

Your tutor is important, especially in warning against falling for trite arguments that can appear beguiling. Watch for moral arguments which are acceptable in one culture or time, but inappropriate in another culture or time: evaluating third-world development polices using the standards and attitudes of a twenty-first-century northern European will not work. Arguing that something 'is the best' is also fraught with danger. There is a good case to argue that Humboldt is the greatest geographer, but that accolade, I might argue, goes to Captain Cook. Some people might want to credit Johnston, Haggett or … (these two sentences could keep a tutorial group arguing for some time). These last few sentences are extreme, unbalanced and unsupported as they stand. Who is the TOP environmentalist or earth scientist? There are many candidates to discuss. Avoid unsupported arguments and strong language unless there is overwhelming evidence.

Ethical issues

All research involving people raises questions around the ethics and rights of the participants and the researcher. Departments and universities have their own guidelines. When working with people, interviewing participants directly or on the telephone, using video or observing group interactions **follow the guidelines**; they may affect the way you structure your research programme. There are ethical issues about confidentiality, obtaining consent to use the material, photographs, publishing, anonymity for respondents and action to take if participants decide later they want to withdraw their comments. The process also protects you from accusations of unethical behaviour.

25.4 Are resources available?

Is the data you need available at the scale and timeframe you want? There may be plenty of Spanish customs documents for trade in the nineteenth century, but if they are locked in a Cadiz office and you do not read Spanish, they are no good to you. If the data does not exist at the right scale and in enough detail, your project will need to be rethought. Do a pilot study.

Organise equipment and software in advance. Check the computer and laboratory opening hours. Book laboratory space and equipment for the days after sampling. One day's idyllic rowing across a lake collecting water samples can require five days' laboratory analysis which may involve organising accommodation and time off work.

25.5 Are pilot surveys worthwhile?

Every lecturer nags every student to do a pilot survey; about 90 per cent do not bother and about 89.9 per cent of these get into a mess as a result. PILOT SURVEYS ARE A GOOD IDEA. They allow you to grab a few samples, trial a questionnaire, run a programme, AND analyse the pilot dataset through all the laboratory, statistical and modelling procedures, putting dots on graphs and numbers into equations. Life is too complicated and the environment much too complex for anyone to get everything right first time. A pilot survey shows there are other angles, other people to consult, suggests other variables to measure, that a screwdriver could be essential – and just being in the field, on site, at a computer or whatever indicates other things you could do to enhance the research. At the very least, go and look at a site before doing oodles of preparation. When working overseas, practise your techniques on comparable local sites before leaving. Make sure you have all the right equipment and a complete checklist of the data you need. At the end of a pilot ask yourself:
? Is this the best approach?
? Can I get all the data needed to answer the hypothesis?
? Can the project be done in the time allowed?
Then revise the methodology.

25.6 What's the format and assessment?

A dissertation or thesis is a formal piece of writing. Lecturers expect you to adhere to the 'rules', make the presentation consistent and tidy, and observe the word length. Ensure that the same typeface is used throughout, figure titles appear in consistent positions, there are page numbers on ALL pages, all the figures and tables are included. Double-check the university guidelines and follow them, especially the WORD LENGTH. Figure 25.2 outlines a typical format with suggested word and page lengths, but many high-quality dissertations follow different combinations. Write with word count in mind – drafting 12,000 words when you need 3,000 is inefficient.

For assessment information find your department's guidelines or use Figure 25.3. There are many variations; some departments ask for an initial plan that counts for 2–5 credits. The per cent distributions are a guide; mark distributions vary considerably depending on the nature of the dissertation.

Chapter	Contents		Page/Word Length
0	Title Page, Acknowledgements, Abstract, Table of Contents, Table of Figures.		Keep short
1	Introduction: Brief background, Research aims, Signpost thesis layout.		2–3 /600–1,000
2	Literature Review: summary of material relevant to this research. Links to geography, environment or earth science issues.		4–8 /2,000–3,000
3	**Theoretical Thesis**	**Empirical Thesis**	2–3 /1,000–2,500
	Methodology: a description of your research approachOR	Methodology: your research process, techniques used, criticism of techniques and evaluation of their accuracy and representativeness.	
	Discussion and evaluation of first Theme/Idea/Concept.	Site information.	
4	Discussion and evaluation of Second Theme/Idea/Concept.	Results: with tabulations, graphs, and maps, as required.	1–2 / 1,000–2000
5	Discussion and evaluation of third Theme/Idea/Concept/Counter-themes.	Discursive interpretation of results, this may include further modelling work.	2–3 / 1,000–2,500
	Synthesis of themes and alternatives.	Evaluation of accuracy, representativeness, and sensitivity analysis if relevant.	
6	Implications for future research. *'What I would have done if I had known at the start what I know now'.*		1 / c400
	Conclusions		1 / c400
Appendix	Data sets if required. Copy of questionnaire, computer programme, sample of interview transcripts.		Minimise

Figure 25.2 Typical dissertation/extended project format; always check the page or word length rules.

25.7 When do I start writing?

'*I'm Writing-up*' implies writing can be done in one go. Developing ideas and seeing the implications of results takes time. Remember to write as you go. GET SOMETHING ON PAPER EVERY WEEK (Figure 25.1). Read a couple of things and then draft some paragraphs for the literature review. Add to it when you read the next paper. Writing is part of the research activity: you read a bit, write a bit, think a bit; and the combination of these three activities tells you what you might do next. Similarly write up what you do every time you do something. Capture the action!

You cannot hope to have read all the past literature and to re-search all aspects of a topic. Great dissertation management involves stopping reading, stopping investigating and starting writing. You are aiming to report on what you have read and discovered. Your opinions are based on cited material (*references*) and your results. Given another year to complete your dissertation, you would still be reporting on partial information. Your examiners want to see that you have tackled a reasonable topic in a relevant manner and drawn sensible conclusions. They do not expect you to model the processes of global warming at the planetary scale, account for the stratigraphy of all Pacific subduction zones, or explain why buses come in threes. Don't get overwhelmed by reading, keep it in balance. Use the checklist in **Try This 4.6** as a guide and remember to balance 20 minutes of online bibliographic searching with at least a couple of hours of reading.

Dissertation Assessment	
Planning Phase	10%
Clarity in formulation of project. Originality in formulation of project. Independent development of project?	
Abstract	5%
Literature Review	15%
Relevance of literature selected? Comprehensive? Critical comments on literature?	
Methodology	20%
Appropriate to topic? Successful in execution? Followed plan and adapted it appropriately?	
Analysis and Interpretation	25%
Planned? Appropriate? Extent to which aims were met? Consciousness of limitations?	
Discussion and Conclusions	15%
Logical and thought through? Sustainability of conclusions? Suggestions for future research?	
Presentation	10%
Quality of figures, tables, maps, and photographs? References? Appendices? Page numbers? Appropriate length?	
Degree of Supervision Required	
Did the student take the initiative? Any illness or personal problems?	
Further Comments	

Figure 25.3 Example dissertation assessment criteria

⇨ TOP TIPS

✓ START WRITING at the start: the draft will not be right the first or
   second time. Ask a tutor how often s/he rewrites a paper before sending it to
   a journal? 'Lots' is the only answer worth believing (Figure 25.1).
✓ Spell and grammar check everything.
✓ References matter: put them in the document from day 1 and double-check
   your references are properly cited. It is miles easier to do this as you go
   along. Getting to the last three days and realising you have no references is
   like getting to six inches from the summit of Everest and having to go back
   to base camp for a flag and camera.
✓ Proofread what you have written, not what you think you wrote. Make sure
   there are titles and keys on graphs and maps, scale bars, units with equations
   and data, and that the title page, abstract, acknowledgements and contents
   page are in the right format for your department. SPELL and GRAMMAR
   CHECK EVERYTHING.
✓ Proofreading is not easy: get a flatmate to read through for grammar, spelling
   and general understanding. If a friend understands your 'Environmental
   Impact Study of Camels in Klatch' then so, probably, will your tutor. You can
   repay the favour by checking out your mate's dissertation on 'The Humour
   of Baldrick'. Use 'Read Aloud' type software to hear what you actually wrote.
   Leave a few days between writing and proofing – you spot more errors.
✓ Abstract: write this last.

25.8 Safety matters

If you break your leg on fieldwork, don't come running to me.

Dissertation research should be good fun but accidents happen; be aware of your
responsibilities and safety procedures. Safety is as important for human geographers in town as for speleologists down a pothole. Each university and department
has its own safety policy, staff provide advice but you are responsible for you. Most
safety involves common sense and taking sensible precautions. Dissertation fieldwork is usually a solo activity, but do not work alone. Offer 'a day-out' opportunity
to friends and relations – two heads are better than one, and the company will keep
you cheerful. It is your responsibility to:
✓ Take and keep medication or special foodstuffs with you at all times.
✓ Provide yourself with clothing, boots or wellingtons that are appropriate for
   your site conditions.

✓ Check weather forecasts.

✓ Leave a plan and map showing where you intend to go with someone who will alert the authorities if you don't return by a set time. Remember to check in with them when you get back; the police and mountain rescue get very unhappy searching for people tucked up in the local pub reliving the delights of the day.

✓ Take a mobile phone to alert your contact of a change of plan or a problem. In remote areas taking a first-aid kit, map, torch, compass, spare food and a flask with a hot drink is sensible.

✓ Take no unnecessary risks.

Generally, fieldwork safety advice is aimed at people heading for remote uninhabited areas of Scotland, Iceland or Tunisia, but working in towns and cities can be as hazardous. Take care on unfamiliar roads, and especially in countries where drivers are on the other side of the road. Be courteous at all times with the public. People are often curious. Have a clear, non-confrontational explanation for your activities and explain why you have chosen their area. Clothing must be appropriate and professional. If you are not fluent in the local language ask someone to write or record a short statement explaining what you are doing so you can inform people. Before research in the mountains or on glaciers, you may be asked to do a mountain safety course. For international work see Nash (2000a, b), Scheyvens (2014) and Robson and Willis (1997), whose advice applies equally to undergraduates and postgraduates.

25.9 The dissertation viva

A viva is part of assessment in some programmes, where you discuss your research with one of the examiners. Remember you are the expert, you did the research and wrote it up. Prepare by rereading your report, think (*make notes*) about what you discovered and be ready for these questions:

? Can you please start by summarising what you did and what the results mean for … (geology, tectonics, urban studies …)?

? What were the three most challenging aspects and how did you handle them?

? What are the two main strengths of your dissertation research?

? Please outline any weaknesses in the research.

? Please explain why you used this methodology.

? Since you started on this topic I see you did a module on … How might you have adapted your research having done this module?

? Please can you explain how your results relate to hydrology/cultural geography/ petrology/limnology/geology/sustainability/environmental chemistry … generally?

? Obviously in a student project there is limited time for data acquisition; how might you have expanded the data collection process?

? Please talk about sources of error in the data.

? What should a student working on this topic next year do to make the most of what you have done and develop the ideas/understanding …?

Keep your answers to the point and keep up the technical content, evidence, examples, and references, just as in an essay. Every examiner recognises waffle having supervised, marked and moderated thousands of dissertations. They are impressed by people who have done the research and realise there were other things to do, other ways to tackle the issue, other techniques, more data to collect … so tell him/her that you know that too. Enjoy.

25.10 References and resources

Clifford N, Cope M, Gillespie TW and French S (Eds.) 2016 *Key Methods in Geography*, (3rd Edn.), Sage Publications Ltd, London.

Coles T, Duval T and Shaw G 2012 *Student's Guide to Writing Dissertations and Theses in Tourism Studies and Related Disciplines*, Routledge, London.

Dawson C 2009 *Introduction to Research Methods: A Practical Guide for Anyone Undertaking a Research Project*, (4th Edn.), How To Books Ltd, Oxford.

Flowerdew R and Martin D (Eds.) 2013 *Methods in Human Geography: a guide for students doing a research project*, (2nd Edn.), Routledge, London.

Geological Society 2017 Mapping Projects and Research Projects, https://www.geolsoc.org.uk/Geology-Career-Pathways/University/During-your-degree/Fieldwork/Undergraduate-Mapping-Projects Accessed 16 January 2019.

Geoverse 2018 Geoverse e-journal of undergraduate research in geography, http://geoverse.brookes.ac.uk/ Accessed 16 January 2019.

Hoskin B, Gill W and Burkill S 2003 Research design for dissertations and projects, in Rogers A and Viles H (Eds.) *The Student's Companion to Geography*, (2nd Edn.), Blackwell, Oxford, 197–203.

Nash DJ 2000a Doing Independent Overseas Fieldwork 1: practicalities and pitfalls, *Journal of Geography in Higher Education*, Directions, 24, 1, 139–149.

Nash DJ 2000b Doing Independent Overseas Fieldwork 2: Getting Funding, *Journal of Geography in Higher Education*, Directions, 24, 3, 425–433.

Parsons T 2015 *How to Do Your Dissertation in Geography and Related Disciplines*, (3rd Edn.), Routledge, London.

Peters K 2017 *Your Human Geography Dissertation*, Sage Publications Ltd, London.

Plymouth Student Scientist 2018 *The Plymouth Student Scientist*, http://bcur.org/journals/index.php/TPSS/index Accessed 16 January 2019.

Reinvention 2018 *Reinvention: a Journal of Undergraduate Research*, https://warwick.ac.uk/fac/cross_fac/iatl/reinvention/ Accessed 16 January 2019.

Robson E and Willis K (Eds.) 1997 *Postgraduate Fieldwork in Developing Areas: a rough guide* (2nd Edn.), Monograph 8, Developing Areas Research Group, Royal Geographical Society – Institute of British Geographers, London.

Scheyvens R (Ed.) 2014 *Development Fieldwork: A Practical Guide*, Sage, London.

Swetnam D and Swetnam R 2010 *Writing Your Dissertation: The Bestselling Guide to Planning, Preparing and Presenting First-Class Work*, (3rd Edn.), How To Books Ltd, Oxford.

Safety

Search 'fieldwork safety' + your University name for guidance; St John Ambulance, St Andrew Ambulance and the British Red Cross have useful apps and First-Aid Manuals; the British Mountaineering Council BMC TV have various videos and booklets at http://www.thebmc.co.uk/.

Garside, J 2010 *Safety on Mountains*, https://www.thebmc.co.uk/safety-on-mountains-booklet?s=3 Accessed 16 January 2019.

Whiteside, J 2010 *Call Out Mountain Rescue?: A Pocket Guide to Safety on the Hill*, (2nd Edn.), Mountain Rescue Council, London.

Add up 3

Work out the total for the top of each pyramid. For answers see Chapter 28.

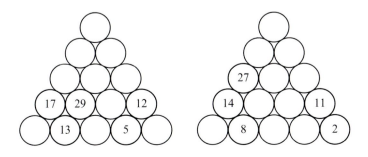

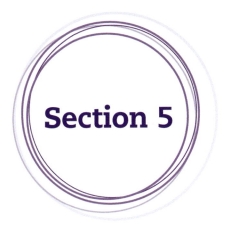

Section 5

Field activities

 Geography, environmental science, geology and related studies are about understanding how the planet works, and how people live and interact with it. Becoming a discipline expert through your degree studies happens because you spend time in the environment developing your understand- ing and appreciating the complexity of the natural environment. Fieldwork is 'where the magic' happens, where your discussions, listening and reading are properly put into context.

 Day visits and longer trips have a learning agenda but are also designed to give you the confidence to plan and run your own projects and expeditions. You will become familiar with practical, ethical and safety matters, and undertake primary research. Expect to be involved in collecting data through experiments, observation, surveys, interviews, and to gain confidence in choosing field techniques and analysis processes in the laboratory and statistically. Chapter 26 covers the practicalities, including safety, which your tutors will emphasise and re-emphasise, while Chapter 27 concentrates on some practical field skills for capturing and reporting data that can be practised at any time.

Geojumble 2

Find the GEES terms in these jumbled letters. Answers in Chapter 28.

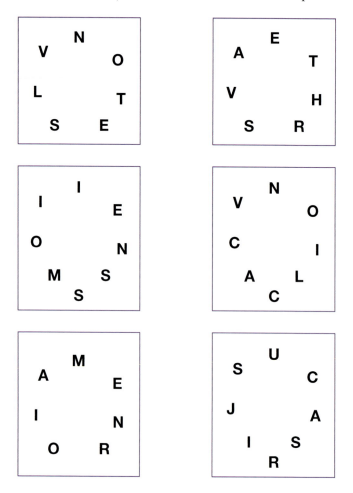

26

Fieldtrips

A geopolitical boundary is an imaginary line between two nations, separating the imaginary rights of one, from the imaginary rights of the other.

Occasionally a student confuses a field-class with a fantastic holiday with mates, but the staff have a well-planned learning agenda for you. University departments use a considerable proportion of their budgets funding fieldwork and field mapping classes because they are convinced it:

☺ develops the observation and research skills that are core to practising geography, environmental science, geology and all the related disciplines;

☺ gives a realistic insight into research activities and practicalities;

☺ allows hands-on, practical experience of field techniques;

☺ encourages individuals to take responsibility for research decisions;

☺ develops group and interpersonal skills (leadership and team skills);

☺ encourages students to talk through research issues with staff in small groups;

☺ allows you to experience some unfamiliar environments, and get the confidence to work in the field in teams and alone;

☺ helps you to develop the skills to manage when 'stuff' goes wrong. ('How do you cope with adversity?', or 'Explain what you did when something went wrong', are common interview questions.)

Your department will provide detailed briefings including costs at open days, in handbooks and online, as well as detailed safety information briefings for every trip. It is usual for students to pay for some of the travel and accommodation costs, especially on longer international field-classes. Find out what is involved.

Fieldwork is a short experience, providing a 'taste' of research activities. The speed of a field-class may mean your technical achievements are limited. At the end of a field visit list the skills you have used and improved, and add them to your reflective log or CV. Skills acquired through fieldwork include organisation, independence, teamwork, measurement, description, safety, use of maps and statistics. Also communication and presentation skills through negotiation and discussion, podcasts, videocasts, posters, maps, field notebooks and reports.

26.1 When is it?

Residential field and mapping courses are usually booked a year in advance. Find out the dates and plan accordingly. First-year field trip dates and costs should be advertised in pre-course literature. Do not agree to skiing holidays, rugby trips or family silver wedding celebrations during this period. Field trips are usually a whole module, or part of a module, and therefore the marks matter.

26.2 Time management

Generally these are easy modules to manage because your lecturers organise you. Expect something like: breakfast at 8, leave at 9, exciting field activities 10–5, dinner at 6, analysis, discussion and report from 7.15, hand in written report at 9 the next day. Field-classes prove you can get up and be ready to work at 9 every day for a week despite having enjoyed the company of colleagues for much of the night. Parents are prone to wonder why you need three days' sleep on return from field-class. Tell them something comforting about sad, sadistic but fit lecturers that made you walk 9 miles each day, AND collect data, AND use computers, AND write it all up before bed AND your room-mate snored – or something similar.

26.3 Preparation

Practical

Six weeks before a field-class check out your passport, injections, holes in wellies, and arrange to borrow rucksacks, wet-weather gear, clipboards and pens and pencils that keep writing no matter how hard it rains. Ask people who went last year what you really need to take, BUT just because last year there was full sun on Denbigh Moor does not mean the temperature can be guaranteed to rise above arctic-survival-bag point this year. Expect staff to want to view the landscape from the top of the hill, a hill rarely equipped with a road or a path. Consider borrowing boots. Geologists should prepare to head into quarries and cuttings that are muddy and on the other side of a hedge, with nettles or gorse bushes.

Many trips are to places without the internet. Download apps, data, and … in advance. The great advantage of working without the distractions of tablets and phones is the time to talk, discuss, listen and plan. By week zero you should be ready.

Academic

Where are you going? It makes sense to look at a map and get some background information, whether it is Bangor or Bangalore. Some departments have pre-departure exercises and briefings, others rather assume that as an intelligent

18+ person you will know that it may snow in the Mediterranean in January and that you need euros to buy drinks in Greece. Ask last year's group what happened. Reminiscing about field-classes over a drink is normal GEES business.

Using Google Earth as an app on your phone or online will show you the landscapes you will visit. Useful for site references and sorting out directions.

There may be virtual fieldwork resources to view and/or information from other classes that have visited the area, try YouTube. Bain (2011) has a brilliant virtual guidebooks site that is worth a visit before any USA trip, for fieldwork or a holiday. Don Bain is a retired geography lecturer from Berkley, California who has created a massive resource of panoramas that allow you to see around the landscape in 360°.

26.4 Using drones

Drone technology is used to capture landscapes from the air, giving new insights into landscape matters. The technology and safety rules are evolving rapidly (Drone-safe 2018), and each country has their own rules. It's essential to check the local rules before flying. In field-class there will be guidance, but if you are using drones as part of your own research make sure you have the right training, and safety clearance from the department. In pilot research a drone can be effective in showing the wider environment and allow risk assessments to be made. When including a photograph taken from a drone in a report, include the elevation in the Figure legend.

26.5 Safety

Someone will alert you to fieldwork safety issues. Universities have safety codes, some will ask you to sign to confirm that you are aware of safety issues, insurance and your responsibilities. The online site by Williams *et al.* (2012) is comprehensive. Generally you will be expected to:

✓ co-operate with the staff about safety and behaviour in the field;
✓ act in such a way that neither you, nor anyone else in the group, is put at risk;
✓ not wander off or do something hazardous;
✓ make staff aware in advance if you have a medical condition (for example: allergies, asthma, diabetes, motor disability), and bring your medication;
✓ know what to do if something goes wrong, such as becoming detached from the group, missing the bus or becoming disoriented if the weather deteriorates and visibility is lost;
✓ bring appropriate footwear and clothing.

Lecturers have completed risk-assessment documentation, planned for problems and gained access permissions. They are usually first-aiders. They will check the weather and adjust activities accordingly.

In rough terrain, field paths, hills, anywhere out of town, walking boots are safest. For river work or on peat bogs, wellingtons with soles that grip are useful. To keep warm, thin layers – T-shirt, shirt, sweater and fleece – under a cagoule are ideal. Layers can be removed, or added, during the day. A woolly hat and gloves are essential for fieldwork in the UK from September to May, and ideally waterproof trousers and a bright cagoule with hood to go over an anorak or fleece. Hard hats will be supplied when needed. Good field gear is expensive; can you borrow clothes for your field-class?

Safety is mostly applied common sense. As a general rule of thumb: 'if you wouldn't … with Grandma or a baby on a Sunday walk', it probably is a very bad idea. On the bad ideas list are: standing in the middle of roads to sketch buildings; trespassing on private property; entering caves, mine workings, working quarries, derelict buildings, or any moving water that is more than 3cm deep. Stick to public footpaths and rights of way. Wear hard hats in quarries, under cliffs and on steep slopes, where rock or debris falls are a hazard. Never climb on cliffs or rock faces without the right kit and supervision. Take care on foreshores, rocky beaches and watch the tide. Do not light fires, and always extinguish matches and cigarettes safely.

Follow your lecturer's advice. Soil and water can carry bacteria, viruses and other pathogens that cause disease. Use appropriate sampling methods and gloves as advised and ensure that you wipe your hands before eating. Everyone on fieldwork should be sensible, polite, considerate of others, and take special care not to damage anything, from hotels to field walls, or to leave gates open.

Take care and have fun, most field-classes are very well planned and free of incidents.

And while away, be environmentally friendly. Use recycled toilet paper.

26.6 References and resources

Search for 'Field trip' and 'Fieldwork safety' – there are many university information sites. See the Safety references, Chapter 25.10.

Bain D 2011 Don Bain's Virtual Guidebooks, http://virtualguidebooks.com/ Accessed 16 January 2019.

Coe AL, Argles TW, Rothery DA and Spicer RA (Eds.) 2010 *Geological Field Techniques*, Student's Companion Site, Wiley Blackwell, London, http://bcs. wiley.com/he-bcs/Books?action=index&bcsId=6048&itemId=1444330624 Accessed 16 January 2019. Dronesafe 2018 Drone code, https://dronesafe.uk/ Accessed 16 January 2019.

Williams A, Williams P and Boyle A 2012 Fieldwork Safety, University of Liverpool, http://pcwww.liv.ac.uk/geo-oer/fs/ Accessed 16 January 2019.

Quick crossword 3

Answers in Chapter 28.

Across

1 Smart conifer (6)
5 Period (4)
8 Young sheep (4)
9 Mine waste (8)
10 Dancer and Prancer (8)
11 Bog fuel (4)
12 Ouse and Trent estuary (6)
14 Running water (6)
16 Cast an eye over (4)
18 … lights (8)
20 Fault, puts thru (anag.) (8)
21 Digital information (4)
22 Scottish Loch (4)
23 Sixth planet (6)

Down

2 Level land (7)
3 Of the town (5)
4 Enterprise initiator (12)
5 Holiday maker (7)
6 Molten rock (5)
7 High cloud (12)
13 Seafloor flora and fauna (7)
15 Used to oxygenate water (7)
17 Small wood (5)
19 Regular behaviour (5)

Presenting maps, sketches and data

With a graphics programme or sketch you can move mountains.

Geography, environmental and earth sciences are visual subjects. Graphics, drawing, plans and mapping are core features. Teaching sessions are usually supported with visual materials because they are very effective ways to convey information. The skill in your assignments is to use visual images, being as professional in production as possible. This means careful checking to make sure each figure has the right level of detail and is accurate. Unfortunately, the author of this chapter has included less than perfect figures (2.2 level work). **There are deliberate mistakes** on all the figures; please exercise your critical skills to spot the mistakes and decide how to improve each one.

Everyone doing science or map work at school learns to add labels, scales, legends, keys and titles because every map and diagram must have them. Lecturers cannot give full marks if they are missing. This is not because life is grossly unfair, which it is, but because geographers, environmental and earth scientists always label all maps and diagrams correctly! The skill here is to give careful, consistent attention to detail and completeness. Don't forget to acknowledge sources with references in figure titles, as in 'Figure x … title … (after Benny and Dibble 2035)'. See also Chapter 19.3 and 19.4 for advice on presenting lab and field notebooks.

Statistics and graphics programmes will provide keys and legends, label axes and format your data. You must be critical and make sure the information presented is exactly what you need. In 99 per cent of cases you will need to make adjustments. Be especially careful of the y-axis which most programmes fit to the maximum data point rather than to a sensible round number.

Remember to use figures in exam answers. Sometimes you will draw in examinations: 'Provide an annotated plan of a new retail park, with brief justification for the features included', 'Draw a geologic map to illustrate sinistral and dextral wrench faults', 'Sketch the micro-depositional forms you find in alluvial channels'. Check past exam papers and practise some answers.

27.1 Field or panorama sketches

Tragically, you will find yourself walking happily in Arran, Abingdon or Algeria when some sad academic says brightly, 'Please do a field sketch of the lithological/

urban/cultural/geomorphological/biogeographical/… features'. Everyone mutters about not being art students. Lecturers have heard this before and will be deeply unimpressed. A rough sketch can convey considerable geological, environmental and geographical evidence, making sketches invaluable in reports and essays. Good lecturers sketch with you so you can see what to do. Figure 27.1 is not particularly good, being part sketch and part cross-sectional plan. There is considerable room for artistic improvement, but it will get good marks because it picks out the **evidence** for slope movement through specific examples. It also shows a 'wider view' than a camera. Five photos would be required to show all these points. This sketch includes evidence the camera cannot 'see' behind cars and trees. That is important.

Great field sketches take practice, but not much. The trick is to pick out the two or three lines that anchor the sketch, usually the horizon, maybe a building or two, and then place the features you want to illustrate. Geologists focusing on rocks, faults and surfaces find that adding a building or a tree helps too (see BGS 2018, Noad 2016, Williams 2016).

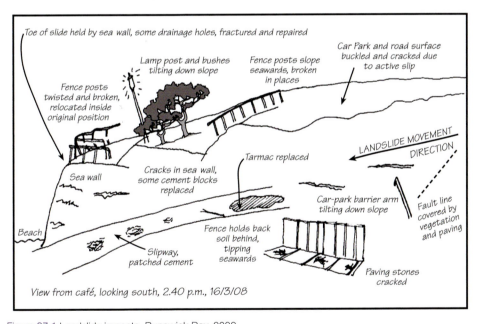

Figure 27.1 Landslide impacts, Runswick Bay, 2008

⇨ TOP TIPS

✓ Draw the features 'that matter for this exercise only'. Attempting to draw the whole landscape is a waste of time (unless you are Constable reincarnated). Keep it simple.

✓ Helpful tools include a firm board, paper, 2B and 4B pencils, pen, eraser and, perhaps, binoculars.

✓ Locate a couple of 'landmarks' to anchor the sketch, then simplify.

✓ Show general outlines for buildings or slopes, omit details.

✓ Shading gives a feeling of depth and distance – turn the pencil sideways. Use thicker, darker lines for close objects and finer lighter lines for those in the distance.

✓ Everyone exaggerates the height of objects like hills and buildings to start with. Try a second sketch reducing the heights by half to two thirds and compare the result. Shadows clutter a sketch; they are a transient element, a function of the weather and time of the visit, and best omitted.

An artist sketching a landscape covers his work with notes on colour, shading and lighting effects. Good field sketches are covered with notes making relevant hydrological/geophysical/cultural/… points. Notes get marks.

First-time sketchers can find it helpful to start with a sky sketch! Skies are generally empty, clouds have simple shapes (Figure 27.2). The only landscape element required is the horizon. Try it from your window now! Minimum labelling involves naming the features, in this case the types of clouds, then adding information on cloud elevations and whether there will be rain soon. Look at Figure 27.2, what other information might be added? Be critical.

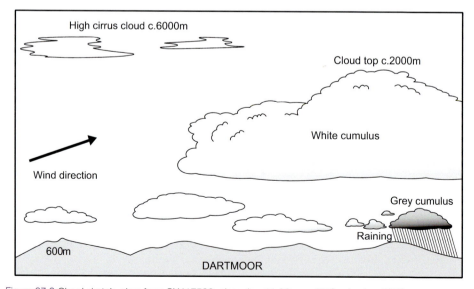

Figure 27.2 Cloud sketch, view from SX417599, clear day, 11.30 am, 20 September 2010

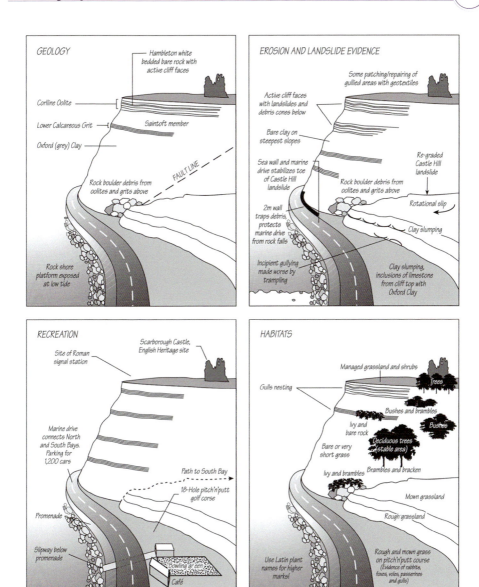

Figure 27.3 Sketches of Castle Hill, Scarborough, UK

Figure 27.3 shows four sketches of the same site, drawn to illustrate four different aspects of the landscape. Pick out the geological/environmental/geographical features you want and ignore the rest. Excellent answers would add a small map to locate the view, the GPS or grid reference of the sketching point and a compass bearing. Make sure your notes are legible.

Figure 27.3 started as sketches which were scanned into a graphics programme and the shading added. They may look more 'professional', but the hand-drawn version is perfect for field reports and dissertations.

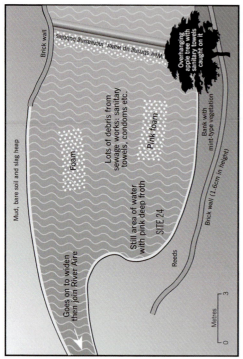

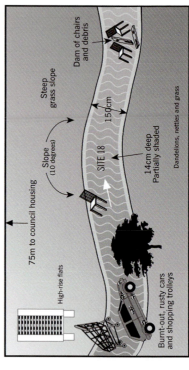

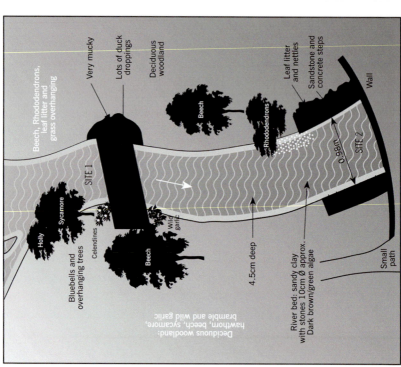

Figure 27.4 Sketch plans of three stream reaches. What is missing from each?

27.2 Field plans and maps

Plotting, sketching or mapping is used to exemplify and clarify spatial relationships. Figure 27.4 shows sketch plans of urban stream reaches. They make multiple points concerning flow, vegetation and debris. An indication of SCALE is vital: this is a small tributary not the Mississippi. One diagram has a scale bar, on the other two the width of the stream is shown. In the final version, replace the plant names with their Latin names.

Sketch maps highlight specific geographical and geological evidence. Either sketch freehand or use a published map as a base, and annotate copiously. Unless there is a very good reason not to, put North at the top of the diagram or map, add a direction arrow and scale bar. Simplification is the key, do not attempt to be comprehensive: convey specific information. 'Anchor' the map with landmarks that will be there ten years later: rivers, railways, power station.

It is logical to use conventional signs from your national map service (UK – Ordnance Survey) and geology maps. Symbols used in geomorphological mapping (Noad 2016, Otto and Smith 2013) are very useful. When working in black and white make sure you can tell the difference between a road, a river and a railway line.

27.3 Tables, graphs and data

Manipulating data to create tables, histograms and graphs is straightforward; inserted into reports, essays and dissertations they look professional and save words. They are not, of course, a substitute for thoughtful analysis. Look very critically at your own and published diagrams, most diagrams can be improved. Tables of data are inherently boring. Consider ways to display data more interestingly, supporting your arguments.

Computer packages will create and colour graphs, bar charts, pie charts … Don't let the output get in the way of your message. Aim for illustrations that are clear, legible, interesting *and* convey your message accurately.

 TOP TIPS

Start by clarifying information for the reader. Use this checklist:
- ✓ Is the data correct? A very high or low data point may indicate a data entry error. Double-check, look back at field notebooks and spreadsheets. Take plenty of time because a data error may totally alter your interpretation of the results.
- ✓ Do graphs have full and explanatory axes labels, units, and titles?
- ✓ Does the presentation design make the information difficult to read and understand?

✓ Is the referencing complete? Does the citation acknowledge the source of the data?

✓ Is it clear whether the data is primary or from another source?

✓ Are there error bars on graphs and histograms?

✓ Do mean, median or modal plots show the data range indicating the precision of the data?

✓ Do regression plots include r values?

✓ Do regression lines fit the data? Check there is a balanced distribution of points above and below the line.

✓ Where there are multiple graphs, are the scales on each graph the same, enabling visual comparisons? This problem arises when computer packages scale the y-axis to fit the range of the data.

27.4 References and resources

Search for *geological mapping*, *geomorphological maps* and *field sketching* to find many examples, images and videos.

BGS 2018 British Geological Survey Map Symbols and Lithological Ornaments, British Geological Survey, http://www.bgs.ac.uk/downloads/start.cfm?id=303 Accessed 15 January 2019.

Noad J 2016 The (Forgotten?) Art of Geological Field Sketches, AAPG (American Association of Petroleum Geologists), http://www.searchanddiscovery.com/pdfz/documents/2016/41853noad/ndx_noad.pdf.html Accessed 15 January 2019.

Otto JC and Smith MJ 2013 Geomorphological mapping, in *Geomorphological Techniques*, British Society for Geomorphology, http://geomorphology.org.uk/sites/default/files/geom_tech_chapters/2.6_GeomMapping.pdf Accessed 15 January 2019.

Williams M 2016 Fieldwork notebook tips, University of Liverpool, http://geohubliverpool.org.uk/field%20sketches.htm Accessed 15 January 2019.

Geo-paths 4

Can you find 15 or more words in the square? Move horizontally and vertically, but not diagonally, from letter to touching letter. Each letter can only be used once in a single word. There is one word that uses all letters. Answers in Chapter 28.

N	T	O	T
E	P	H	O
C	N	I	L
S	E	M	U

28

Answers

Section 1 Wordsearch 1 p 3

F	R	I	E	N	D	S	D	O	O	W
T	G	B	R	R	L	D	I	G	E	O
P	E	V	A	E	H	E	E	S	N	R
E	O	A	A	F	S	O	V	L	A	K
C	I	N	B	C	G	E	I	A	R	O
N	O	L	I	R	A	N	A	T	R	U
O	G	N	A	D	E	T	B	R	L	I
C	Y	P	T	F	D	A	I	V	C	A
S	H	I	C	O	E	Q	K	O	E	H
Y	L	P	W	H	U	C	R	D	N	X
T	N	E	M	N	O	R	I	V	N	E

Chapter 1 Geo-paths 1 p 20

Cine, environ, environment, iron, men, micro, microenvironment, mint, mine, nor, roe, vie, vino.

Chapter 2 Quick crossword 1 p 32

Across: 1 Before; 4 Tuition; 6 Ability; 8 Mirror.
Down: 1 Blurb; 2 Fit; 3 Eon; 5 Outer; 6 Aim; 7 Lor.

Chapter 3 Drop out 1 p 44

Mexican: mammoth, Permian, gumtree, plateau, isotope, geodesy.

Chapter 4 Words in geo-words 1 p 58

There are over 100 words in SEDIMENT. Thirty is a good score; some words can appear twice in the list with and without a final 's'. Well done! You may have found:

Deem, deism, deme, dement, demise, den, denies, denim, dense, dent, diet, din, dine, dies, diet, dint, edit, emend, ends, inset, item, meet, men, mend, mete, mid, midst, mind, mindset, mine, mint, mist, mite, nest, nets, seed, seem, semen, sent, semi, side, sine, site, sited, smite, snide, stem, teem, teen, tee, ten, tend, tense, tide, tied, time, timed, tines, tin.

Chapter 5 Missing vowels 1 p 73

Biodegradable; Converter; Decompose; Landfill; Methane; Packaging; Recovery; Reuse; Separation; Waste Disposal.

Chapter 6 Geojumble 1 p 88

Cyclone, Lignite, Tsunami, Drought, Species, Habitat.

Section 2 Relaxed maths 1 p 90

73; 39; 38; 85; 104; 43; 84; 567; 23; 51; 195; 32.

Chapter 7 Geo-codeword 1 p 98

1	2	3	4	5	6	7	8	9	10	11	12	13
L	P	E	S	D	Z	U	B	X	I	T	W	J

14	15	16	17	18	19	20	21	22	23	24	25	26
G	O	N	A	Q	C	R	M	F	Y	K	V	H

Chapter 8 Geo-paths 2 p 112

Air, ale, are, arrest, arse, art, ate, axe, axel, err, errs, extra, extraterrestrial, lair, lax, let, rare, rat, rate, rest, ria, serrate, tar, tare, tars, tart, tax, taxa, terrestrial, text, trial.

Chapter 9 Geogram pairing 1 p 120

Edges sedge; granite tearing; hearty earthy; leash shale; point piton; sample maples; spinier inspire; trace carte; turbine tribune; unclear nuclear.

Chapter 10 Add up 1 p 129

The totals are 127 and 131.

Chapter 11 Wordsearch 2 p 139

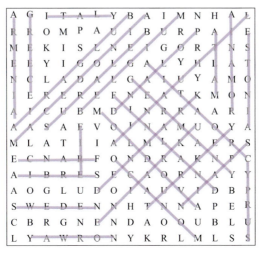

Chapter 12 Geo-codeword 2 p 152

1	2	3	4	5	6	7	8	9	10	11	12	13
G	P	U	E	N	R	C	Y	I	W	T	X	D
14	15	16	17	18	19	20	21	22	23	24	25	26
K	O	B	Z	H	M	S	A	Q	F	J	V	L

Chapter 13 Missing vowels 2 p 167

Air Quality; Biomass; Chaos; Dilute; Erosion; Irrigation; Fertilizer; Groundwater; Ozone; Radioactive.

Chapter 14 Relaxed maths 2 p 178

70; 40; 24; 115; 109; 39; 8; 176; 32; 90; 189; 67.

Chapter 15 Geo-paths 3 p 190

Enter, entrepreneurship, epi, ere, hint, hip, hue, hunt, hunter, net, nip, pen, per, pert, pine, pint, pre, rent, renter, run, rune, runt, shin, shine, shiner, ship, shun, shunt, shunter, ten, urn.

Chapter 16 Words in geo-words 2 p 202

There are over 100 words in PALEONTOLOGY. Thirty is a great score. Well done. You may have found:

Aeon, agent, age, aglet, ago, agony, ale, allot, allotype, alloy, ally, aloe, alone, alp, alto, angel, angle, ant, ante, apology, aptly, atoll, atone, aye, eat, ego, élan, eon, gale, galena, galleon, galley, gallop, gap, gate, gelato, genoa, gently, gloopy, gnat, goal, goat, gooey, lag, lagoon, lane, lap, lapel, lately, lean, leapt, leg, legal, legato, lento, logo, logotype, lonely, looney, loopy, loyal, lye, nag, nape, neap, neatly, neology, noel, oaten, ontology, oology, open, openly, opt, pale, pallet, pan, pat, pea, pen, penalty, plant, planet, planetology, plate, platoon, plenty, ply, poet, polenta, pole, polyglot, polygonal, polytonal, polo, poly, pony, pool, tag, tale, tall, tan, tangelo, tangle, tango, tape, tea, teal, tell, ten, toga, toil, tonally, tool, topology, toy, type, yang, yap, yea, yell, yelp, yet, yoga.

Chapter 17 Quick crossword 2 p 214

Across: 1 Tariffs; 5 Datable; 9 Fecundity; 10 Coypu; 11 Climatologist; 13 Icefield; 15 Linear; 17 Talkie; 19 Diaspora; 22 Rubber stamped; 25 Aorta; 26 Acidifier; 27 Diluted; 28 Mayweed.

Down: 1 Tufa; 2 Recycle; 3 Fungi; 4 Spiracle; 5 Dry rot; 6 Tectonics; 7 Bayside; 8 Equatorial; 12 Hinterland; 14 Itinerant; 16 Titanium; 18 Liberal; 20 Old Time; 21 Island; 23 Privy; 24 Grid.

Chapter 18 Add up 2 p 237

The totals are 115 and 103.

Chapter 19 Be aware of oxymorons p 252

Act naturally, authentic replica, found missing, almost exactly, pretty ugly, larger half, only choice, working holiday, clearly confused, old news, awfully nice.

Chapter 20 Words in geo-words 3 p 255

There are over 80 words in FIELDWORK, with …er and …ed variations. You may have found:

> deli, dew, die, dike, dire, dirk, doe, doer, dowel, dower, drew, fie, field, file, filo, fiord, fir, fire, fled, flew, flier, floe, florid, flow, flowed, flower, foil, folder, folk, for, ford, fried, idle, idol, ilk, irk, kef, kid, kilo, led, lewd, lido, lied, life, like, lire, lode, lord, lore, low, lower, oiled, older, ole, ore, oriel, owe, owed, owl, red, redo, refold, rid, ride, rife, rifle, rile, rod, rode, roe, role, rolled row, rowed, rowel, wed, weir, weird, weirdo, weld, wide, wield, wife, wired, woe, wok, woke, wolf, word, wore, work, world.

Chapter 21 Geogram pairing 2 p 258

Boatel oblate; clast talcs; doable albedo; inhale haline; nacre crane; source course; stream master; tablet battle; teles sleet; trifle filter.

Chapter 22 Wordsearch 3 30 countries p 267

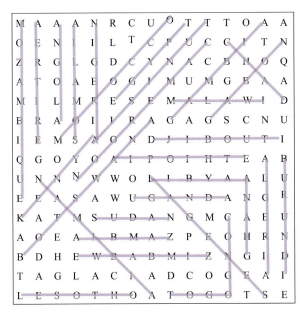

Chapter 23 Relaxed maths 3 p 275

100; 176; 75; 222; 660; 9; 740; 632; 648; 56; 9; 229.

Chapter 24 Wordsearch 4 p 293

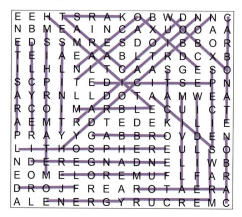

Chapter 25 Add up 3 p 309

The totals are 179 and 106.

Section 5 Geojumble 2 p 312

Solvent, Harvest, Emission, Volcanic, Moraine, Jurassic.

Chapter 26 Quick crossword 3 p 317

Across: 1 Spruce; 5 Time; 8 Lamb; 9 Tailings; 10 Reindeer; 11 Peat; 12 Humber; 14 Stream; 16 Scan; 18 Northern; 20 Upthrust; 21 Byte; 22 Ness; 23 Saturn.
Down: 2 Plateau; 3 Urban; 4 Entrepreneur; 5 Tripper; 6 Magma; 7 Cirrostratus; 13 Benthos; 15 Aerator; 17 Copse; 19 Habit.

Chapter 27 Geo-paths 4 p 325

Cent, cep, emu, him, hot, incent, line, lines, lime, limes, lot, lumen, men, mil, milo, min, mince, mine, mines, nil, pec, pecs, pen, pent, phi, photo, photoluminescent, scent, semi, tot, toto.

Index